Fundamentals of Chaos and Fractals for Cardiology

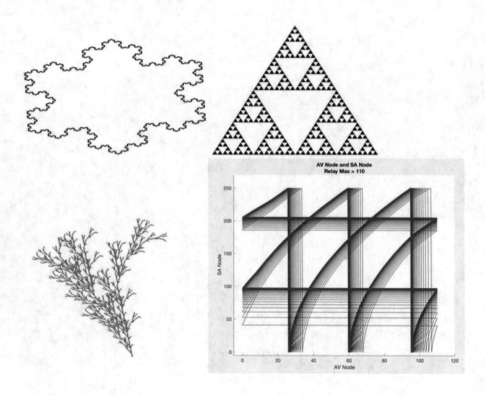

Gary Drzewiecki

Fundamentals of Chaos and Fractals for Cardiology

 Springer

Gary Drzewiecki
Rutgers University
Piscataway, NJ, USA

ISBN 978-3-030-88970-8   ISBN 978-3-030-88968-5   (eBook)
https://doi.org/10.1007/978-3-030-88968-5

This Springer imprint is published by the registered company Springer Nature Switzerland AG
The registered company address is: Gewerbestrasse 11, 6330 Cham, Switzerland

Preface

This book was designed with the physiologist, medical researcher, and biomedical engineering researcher in mind who are not trained in the area of nonlinear dynamics. These same researchers may have seen references to Chaos and Fractals and believe that it might benefit their work. This book will provide that researcher with the background to get started in applying nonlinear dynamics to their work by providing the key fundamentals needed to understand this field. The focus will therefore be on providing the reader with the essential nonlinear tools and definitions needed to make your own decision in applying nonlinear dynamics to their work. The approach to the content of this book is to provide the reader with fundamental rules of Chaos and Fractals in the field. Moreover, the book provides background to the historical origin of Chaos and Fractals. You will also find references to the key authors of Chaos and Fractals. This approach is beneficial in helping the reader to understand new research and developments more easily in the field. Also, if more background is found necessary after reading this book, you will have the key references to find the original sources more easily. Following the presentation of these fundamentals, this book will end with two problems that apply the fundamentals to solve them using first principles.

The straight line has been a chain on humanity—Andrea

The subjects of Chaos and Fractals may be a field unto themselves. The reader may wish to study in more detail in which case you will have learned the fundamentals here. This book has been written for life scientists with a good knowledge of calculus. Prior training in ordinary differential equations would be beneficial as well as some concept of numerical methods.

The applications of Chaos and Fractals are broadening since their beginnings. Nonlinear dynamics has been applied to economics, epidemiology, plant biology, human physiology, and astronomy. The topics of this book are restricted to subjects that demonstrate the first principles of Chaos and Fractals by example. Learning the first principles will enable the reader to more easily proceed to other applications of Chaos and Fractals. As you move through this book, you will be provided with several useful tools that have emerged from nonlinear dynamics that you can immediately apply to your research. Much can be learned from examples of Chaos and Fractals in cardiology. As such, many of the applications of Chaos and Fractals in this book will highlight problems in cardiology and cardiovascular physiology.

Acknowledgments

This book could not be possible without the following people. I dedicate this effort to them.

My wife Claire

My family, the Drzewiecki's

My advisor, from the University of Pennsylvania, Dr. Abraham Noordergraaf

Contents

Introduction

1

1.1 Nonlinear Cardiovascular System Examples

Before moving on to the main content of the book, a few cardiovascular examples are provided so as to further motivate the study of Chaos and Fractals.

1.1.1 Blood Vessel Mechanics

Historically, researchers have used linear approximations to model physiological systems. In most cases, this method has not presented a limitation to physiological modeling since the first-order effects are represented adequately. However, the Chaos response will not be represented since nonlinear properties are required to observe a chaotic response. Blood vessel mechanics provides such an example. A large difference between the results obtained via linear approximation versus a complete solution of the nonlinear system occurs when not incorporating physiological vessel mechanics. Consider the mechanics of a human brachial artery from the perspective of its lumen area versus pressure relationship shown in Fig. 1.1. This figure clearly demonstrates a nonlinear behavior, but to a large extent the vessel is typically modeled as a constant compliance. Since the compliance is the slope of the curve in Fig. 1.1, this consequently is not true. The linear compliance representation of a blood vessel prevents any Chaos response from being observed by accurately incorporating this effect in any analysis. Clearly, linear approximation of the "real" arterial function would be limited to conditions where the normal physiology is restrained to a narrow range of function. In this example, a blood vessel can easily exhibit nonlinear function whenever the external stresses exceed that of the internal pressure. This can occur when an intramuscular vessel, such as the heart, is subjected to muscle contraction.

© The Author(s), under exclusive license to Springer Nature Switzerland AG 2021
G. Drzewiecki, *Fundamentals of Chaos and Fractals for Cardiology*,
https://doi.org/10.1007/978-3-030-88968-5_1

1.1.2   Cardiac Valve

Another example of important nonlinear function in the cardiovascular system is the cardiac valve. The cardiac valve responds to the pressure across it by opening or closing to allow or prevent blood flow. A positive pressure across the valve results in aortic flow. The valve therefore functions as a highly nonlinear blood flow resistance. Without the valve, the mechanics of cardiac pumping would render a very inefficient process at moving blood in a single direction. The need for nonlinear function is certain in the case of the valve. A hypothetical valve pressure–flow function is shown in Fig. 1.2.

Notice that blood flow is only permitted for positive values of valve pressure drop. This creates a flow discontinuity about the zero pressure as well as a nonlinear function for positive pressures. In the case of the valve, we find that its nonlinear

Fig. 1.1 The lumen area of the human brachial artery shown as a function of arterial transmural pressure. (Reproduced from Drzewiecki & Pilla, 1998)

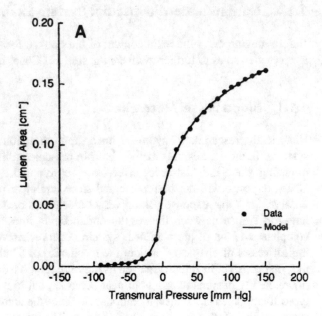

Fig. 1.2 A sample pressure–flow relationship for a human aortic valve. The vertical axis, y, is pressure drop (mmHg) across the valve, and the horizontal axis, q, is blood flow (cm³/sec)

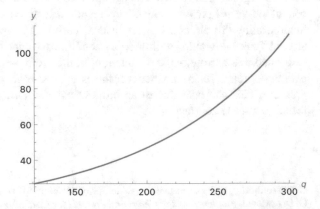

function is an essential function of the cardiovascular system. Other dynamics that arise from valve function are the heart sounds. The first and second heart sounds due to the closure of the aortic and mitral valve are familiar to all cardiologists and are a result of this nonlinear function. The full nonlinear impact of valvular function remains to be studied regarding arterial blood pressure dynamics. It should be clear at this point that nonlinear function is an essential element for Chaos to occur. In the chapters that follow, it will be shown how nonlinear function leads to Chaos in a system cardiovascular or otherwise. In physiology, Chaos will most often occur in the variables that we observe, for example, in time signals. This is detected as a variability in the signal. It might be assumed that the variation is due to a random process. For example, the variation of a pure random process is shown in Fig. 1.3.

1.1.3   Chaotic Signals

Alternative to the random variation shown in Fig. 1.3, consider the variation of a nonlinear dynamic system that is generating chaos as shown in Fig. 1.4.

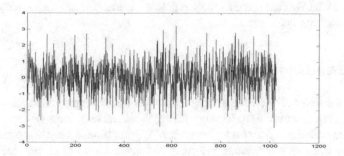

Fig. 1.3 The variation in a variable that is random or pure noise. Time is the horizontal axis. The data were generated using a computational random number generator function

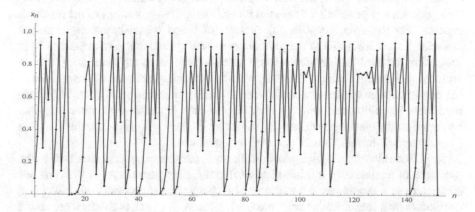

Fig. 1.4 Single variable obtained from a chaotic nonlinear dynamic system for variable x_n versus time point n. (Generated using Mathematica's Wolfram Demonstration Project, Wolfram.com)

The Chaos signal in Fig. 1.4 is a distinctly different kind of variability than that of the random signal in Fig. 1.3. The random signal is truly random and unpredictable. On the other hand, in this example, chaos was generated by a known system. Hence, each point leads to the next in a deterministic way. Hence, we find a clear value to our understanding of Chaos and nonlinear systems in that they are predictable if the system that generated the Chaos is known at least in the short term. Our conclusion from the random variable versus the chaos variable is quite different. In the case of Chaos, the system may be defined. Referring again to Fig. 1.4, Chaos, look for patterns in the time series. For example, you may find repeated values or segments. This might be reasonable to expect since the data in Fig. 1.4 were generated by a computer implementing an algorithm in a repeated manner. These repeated time series patterns may be treated in a more organized way. That is the concept of a Fractal. This leads to the second main topic of this book, which is the Fractal.

1.2   Fractals

As we have seen above, a time series may appear to have repeated values in it even though it is Chaos. This same kind of repetition is also found in geometry, particularly in branching geometries.

1.2.1   Branching Structures

To begin the study of the Fractal concept, it will be necessary to examine repeated structures such as branching. Biology favors the iterative process or repetition such that a more complex system may be achieved. Branching is very common to much of biology since it is the usual case that biology tends to duplicate one function repeatedly. This repetition or iteration can yield a very complex structure over time. It is referred to here as a class of nonlinear function because it will be seen that most iterative procedures yield complex nonlinear functions. Despite the resulting complex process, it is generally possible to identify simplifying features of the resulting process. For example, consider the process of blood vessel branching. Another branching structure is that of the pulmonary airway system that can be created from branching the trachea right down to the alveoli. Other organ systems may be constructed in a similar fashion. For example, the branching structure of the arteries in the cardiovascular system can be viewed as a similar branching structure. We recognize these branching structures as Fractal geometry. Branching structure may also be found in the electrophysiology of the heart. Referring to Fig. 1.5, the Purkinje fiber system of the heart's activation system is shown.

Repetition also recalls biological oscillations, such as the pacemaker. Due to the presence of nonlinearity in biological oscillators, they possess properties that are distinct from the man-made oscillator. The characteristic properties of biological oscillators will therefore be presented. Man-made physical oscillators are much more rigid in the sense that they are designed to follow a math function such as the

Fig. 1.5 Example of a human physiological structure created by repeated branching is demonstrated by the Purkinje fiber neural system of the heart. (Reproduced from Goldberger & West, 1987)

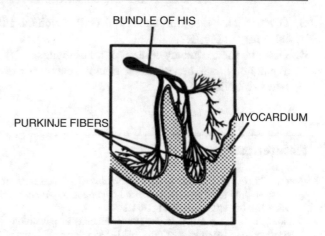

BUNDLE OF HIS

PURKINJE FIBERS

MYOCARDIUM

sinewave. The pacemaker of the heart is of primary interest to the cardiologist. So, the cardiac pacemaker will be reviewed as a nonlinear oscillator in this book. This topic will be presented in the nonlinear cardiology chapter following this introduction to some of the topics of this book.

1.3   Book Organization

This book has been organized into three main parts. Part I will introduce you to the fundamentals of nonlinear dynamic systems and Chaos as well as methods to characterize the nonlinear system response. Part II is devoted to an introduction to Fractals. The Fractal will be introduced by way of using the concept to analyze fractal objects and fractal time. In addition, the deterministic Fractal will be introduced to provide a more mathematical foundation. Part III is devoted to applications of what the reader has learned in Parts I and II primarily to the area of cardiovascular dynamic research. Part III will show by example how to develop and analyze a nonlinear dynamic model. Nonlinear methods will then be used to study the model data. Lastly, this section will illustrate how the tools provided in this book can help to eliminate assumptions and solve problems of cardiovascular physiology that have otherwise gone unsolved using linear methods. The primary emphasis of the book will be to supply the reader with first principles of Chaos and Fractals. This approach will enable an understanding of the current research and the ability to solve research problems.

Problems

1. List five human physiological systems that are nonlinear.
2. Sketch the nonlinear input/output function for each of the systems listed in Problem 1.

3. Given that arterial compliance is dA/dP, sketch the arterial compliance curve for the artery in Fig. 1.1.
4. Describe the frequency content of random noise.
5. What functional benefit does fractal geometry provide to the Purkinje fiber network of the heart?

References

Drzewiecki, G., & Pilla, J. J. (1998). Noninvasive measurement of the human brachial artery pressure-area relation in collapse and hypertension. *Annals of Biomedical Engineering, 26*(6), 965–974. https://doi.org/10.1114/1.130

Goldberger, A. L., & West, B. J. (1987). Fractals in physiology and medicine. *Yale Journal of Biology and Medicine, 60*(5), 421–435. Retrieved from https://www.ncbi.nlm.nih.gov/pubmed/3424875

Part I

Fundamentals of Nonlinear Dynamics and Chaos in Cardiology

Topics in Nonlinear Dynamics

2

2.1 The Cardiovascular System

Applications of nonlinear dynamics to the cardiovascular system are typically related to the electrophysiology of the heart such that the phenomenon of interest is usually events that are of the time intervals less than 1 min. Of particular interest is the heart rate variation and arrhythmias. Nonlinear methods provide a different perspective into electrophysiology and are typically not taught in the classical medical physiology textbook. In particular, physiology can explain and identify the origin of some arrhythmias, but the most common, the sinus arrhythmia, remains to be explained. The cardiac rhythm was generally introduced in Chap. 1 as the concept of the bio-oscillator. A nonlinear oscillator may be used to analyze many known cardiac rhythms normal and pathological. Moreover, since the cardiac pacemaker is a repetitive process, the concept of the Fractal can be employed. Cardiac electrophysiology has enjoyed a long history of research, so that information is available at the heart level, the cellular level, and the molecular level. Nonlinear dynamics will be employed in this book to revisit each structural level with a new vantage point. A focal point of debate is the origin of cardiac fibrillation. Briefly, fibrillation is known as an uncoordinated electrical and mechanical activation of the cardiac muscle that leads to death due to net zero pumping of blood. Some dynamicists have theorized that fibrillation is Chaos. These discussions will be covered in more detail in later chapters after the reader becomes more familiar with nonlinear dynamics. Continuing along with our survey of nonlinear dynamics in physiology, move next to the pulmonary system.

2.2 Respiratory Physiology

Recall from Chap. 1 that the Fractal concept applies well to the modeling of the geometry of the respiratory system. Also, the control of breathing lends itself well to the application of nonlinear dynamic methods. Of course, everyone is familiar with breathing being a periodic process. It would be simple to suggest that breathing is stimulated via pacemaker cells just as in the heart. Surprisingly, the respiratory system possesses no such respiratory oscillator. Instead, the use of a nonlinear feedback control system explains the source of oscillations (Glass & Mackey, 1988). Physiologists are familiar with the homeostatic reflex regulation of breathing. Reflex regulation is a negative feedback control system that is employed extensively in human physiology. By adding nonlinear affecters and feedback delay, it is possible to cause the respiratory regulation system to become unstable and oscillate. This kind of treatment of the respiratory system explains the source of oscillatory breathing. This approach further explains some pathology of breathing such as respiratory arrest and Cheyne's stokes phenomenon and other abnormal breathing patterns.

2.3 Population Dynamics

Probably one of the first applications of nonlinear dynamics to biology is that of the study of population dynamics, that is, to analyze how various animal populations vary in time. Of recent interest is the study of how diseases may spread through a population. The analysis of disease spread is a subject of epidemiology population dynamics analysis applied well in the case of a confined population. For example, a human population that is constrained within the borders of a country. Many readers are familiar with epidemiologist presentations of the spread of COVID infections during 2020 (Giordano et al., 2020). The theory behind the infection–time curves rests on the analysis of population dynamics. Additionally, population dynamics may also be applied to populations of cells. For example, the dynamics of the human blood cell population. Human cell dynamics is useful to help analyze various types of anemia in white and red cell concentration. The analytical knowledge of these cell populations can be used for more effective treatment of disease that causes extreme low-cell concentrations.

In summary, nonlinear dynamics analysis applies well to much of human physiology. We will explore more of these applications as we move through this book. At this time, it is useful to examine some of the analytical methods that are valuable in our study such as basic population models.

2.4 Population Model Solution

To illustrate some of the fundamental methods by which the nonlinear biological system may be analyzed, a simple population model is now introduced and solved. A characteristic feature of unrestrained population growth is that the rate of growth is proportionate to the magnitude of the population itself. Mathematically, this can be written as

$$P^{'}(t) = aP(t) \tag{2.1}$$

where $P(t)$ is the population versus time, and P' is its time derivative; "a" is a constant of proportionality that must be calibrated to the population. The solution of this differential equation is easily determined to be

$$P = PP_0 e^{at} \qquad (2.2)$$

The constant P_0 is the initial population. Following this model, it is found that the population is unstable, that is, as t increases, the population increases exponentially to infinity for $a > 0$. Reexamining the model for the case where $a < 0$, it is found that the model would predict that population decreases exponentially to zero as t increases. Hence, for either sign of the model, results are unreasonable. A sample population growth curve is shown in Fig. 2.1 for a positive value of a, where infinite population growth occurs.

This model is flawed in the final value prediction of the model. There are only two final values: infinite population or a zero population. The model is flawed in these outcomes, but the model predictions are not unreasonable over short time intervals. So, rather than abandoning the model completely, look at a modification. Returning to Eq. 2.1 of the model, it is a linear model that permits the population to grow in direct proportion to its size. This problem can be repaired by adding a term that prevents growth when the population becomes too large. So, modify the population model now to limit growth according to a new differential equation,

$$\frac{dP}{dt} = aP(L - P) \qquad (2.3)$$

where L is the population limit. Here, we will use a normal value of $L = 1$. For Eq. 2.3, it can be seen that as the population grows larger than L, the derivative of population growth will turn negative and prevent further growth. Or, when $L = P$. $dP/dt = 0$, so that the steady state P of this new model is for population to converge to L. Let's confirm this by solving Eq. 2.3 for $P(t)$. At this point, we run into a new problem. That is, after expanding the right side of Eq. 2.3 we see that a term aP^2

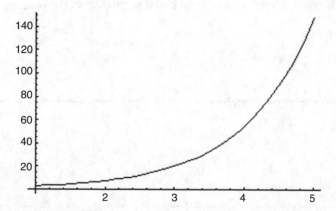

Fig. 2.1 Example of bounded population growth model with unlimited growth. (Generated using Mathematica's Wolfram Demonstration Project, Wolfram.com)

emerges to result in a nonlinear differential equation. Hence, the reader now sees our first example of a nonlinear dynamic system in this book. There is much to learn from this very simple nonlinear model. To solve this equation, Euler method is now introduced. This will be a computational solution of the differential equation.

Start by defining two adjacent population points in time: population at the current time is $P(t_i)$, then, after a small change in time Δt. $P_{i+1} = P(t_i + \Delta t)$. Assuming that the time change can be made infinitely small, the derivative of population then can be written as

$$\frac{dP}{dt} = \frac{P_{i+1} - P_i}{\Delta t}$$

Using the definitions and substituting them into Eq. 2.3, the new model is converted to Euler computational form to be Eq. 2.4:

$$P_{i+1} = \Delta t a P_i \left(1 - P_i\right) + P_i \tag{2.4}$$

This form of our new population model is known to those who study nonlinear dynamics or population modeling as the logistic equation [REF]. Equation 2.4 is now in the form such that we enter the current population value P_i and simply compute the next value of population P_{i+1} Hence, we now have an iterative solution for the original differential equation model Eq. 2.4 A sample iteration of Eq. 2.4 is provided in Fig. 2.2. The properties of this equation are very illustrative of nonlinear dynamic systems. These features will be examined in more detail in a later chapter, but for now notice that the new population model is functioning as required. In Fig. 2.2, the vertical axis represents population, and the horizontal axis is time, where each value n indicates a time step of Δt. Observe that population now converges to a specific level of about 0.5. As you 'll find later, no matter what the initial population starts at, this model will converge to the .5 level. This was determined by the choice of parameter $a = 2$.

The vertical axes represent population value and the horizontal axes represent time. In this graph, a value of $a = 2$ and initial population of 0.6 were used.

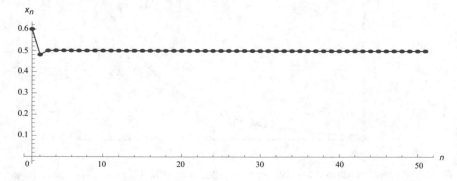

Fig. 2.2 Sample iteration of the logistic equation Eq. 2.4 with $a = 2$

As you study this simple nonlinear model, it is worth noting that the computational form of the model is very simple when it is presented as an iteration as above. Consider that the model constants may be combined, and that the resulting solution is $P(t+\Delta t) = \alpha P(t)$ with all constants combined to be α or as $P_{i+1} = \alpha P_i$. It is clear that the nonlinear system solution has been reduced to a very elementary form. This approach is what biology has been optimized to perform very well where one very simple process is then performed repeatedly to yield a much more complex function. This simplification procedure will be highlighted throughout this book. Thus, the student of nonlinear dynamics need not learn a diverse set of procedures to study the nonlinear dynamic biological systems. The analytical tools that you will learn here will have general application. We have seen that an important biodynamic process is the oscillator. Or, otherwise known as the pacemaker. Hence, it is useful to introduce bio-oscillations into this chapter as a special nonlinear process.

2.5   Bio-oscillation

Bio-oscillators are created from self-excitable cells such as the cardiac sinoatrial and atrioventricular nodes. These groups of cells periodically vary their cell membrane voltage according to a nonlinear dynamic process.

It is important to make clear that the bio-oscillator is quite different from the linear physical or engineering oscillator and is generally unfamiliar to an engineering student. The following is a review of both oscillators to make this clear. Beginning with the classic engineering oscillator, a general oscillatory system is illustrated in Fig. 2.3.

The classic engineering oscillator consists of a feedback system where the feedback is positive and the return feedback is passed through a transfer function that is designed to resonate at the desired frequency of oscillation and cause a total of 360° phase shift. The system will typically lose some energy. So, an amplifier must be included so as to return just enough energy to maintain oscillations. The respiratory system oscillator is a system of this form as mentioned in Chap. 1. If the feedback system is linear, the oscillations will furthermore be sinusoidal in form. Moreover, since the phase shift element is typically a resonant element, the purity of the sinusoidal oscillations can be designed to be nearly perfect in sinusoidal form as required in engineering. As a result of linearity, if this type of oscillator is disturbed, it will

Fig. 2.3 The engineering oscillatory feedback system

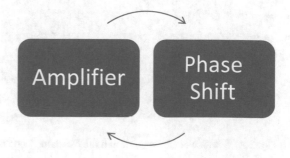

return to its original sinusoidal oscillations prior to the disturbance. This kind of response to an external disturbance differs from that of the function of a nonlinear oscillator. For example, the cardiac pacemaker of the heart drives the atrioventricular node of the heart. Thereby, the SA node serves as the primary rhythm of the heart so that the two oscillators function in a master–slave configuration, where the AV node takes on the frequency of the SA node. The linear feedback oscillator system, such as described above, would not permit this kind of function. Instead, the two pacemakers would simply maintain each frequency independent of each other (Goldberger & West, 1987). Thus, a different kind of oscillator must be used to model the cardiac pacemaker. It is next necessary to introduce a nonlinear mechanism of oscillation that is employed by single cells to cause self-excitation.

2.6   Basic Nonlinear Oscillator

The simplest form of nonlinear oscillator is the integrate and fire model. This type of oscillator functions in a similar way as the self-excitable cell. The integrate and fire oscillator also finds its way into electronic oscillators where it is known as the reflex oscillator (Crisp, 2019). The function of an integrate–fire oscillator must be described in a discontinuous way. For example, let the output of an integrate and fire oscillator be y. This output will vary according to the following: A is a constant. When y is less than the low value L, the value of y increases according to the integration. The opposite occurs when y exceeds the high value of H. This logic causes the value of y to oscillate between the range of L and H:

$$\text{If} : y < L \; y = \int + A dt,$$

$$\text{If} : y > h \; y = \int - A dt$$

Implementing this model yields the result of Fig. 2.4. The model parameters for this example were $A = 7$, $h = 10$, and $L = 0$.

The integrate and fire oscillator corresponds with physiology of self-excited oscillation at the cellular level. For example, if a pacemaker cell is modeled. The

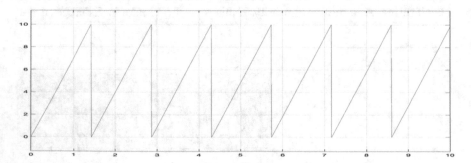

Fig. 2.4 Solution of the integrate and fire oscillator. Time is on the horizontal axis in seconds

cellular membrane possesses the electrical property of a capacitor. Thus, the ionic currents of the cell membrane will be integrated over time in the membrane capacitance to yield a voltage according to the voltage–charge law of a capacitor. The sodium and potassium ion channels are voltage-operated channels and allow a charge discharge current. This corresponds to the If logic of the integrate–fire model. The ionic currents, together with the membrane capacitance, determine the rise and fall times of the membrane voltage and together add to become the pacemaker time period of oscillation. Hence, the pacemaker frequency may be easily modified by external currents that modify the total ion channel current. This allows the pacemaker frequency to be altered by external regulatory signals such that heart rate can be controlled, which is the desirable property of this nonlinear oscillator.

The integrate and fire oscillator model not only applies to physiological oscillation but also the model applies well to common electronic oscillators where it is referred to as the relaxation oscillator. For example, an operational amplifier oscillator circuit is provided in Fig. 2.5. The amplifier serves to detect the high- or low-specific threshold voltage and causes the output voltage to change the direction of charge discharge of the capacitor. A computed capacitor voltage is also included for this circuit in Fig. 2.5. Notice the similarity between the capacitor voltage waveform and that of Fig. 2.4 of the mathematical integrate and fire waveform that illustrates the integration operation in both models.

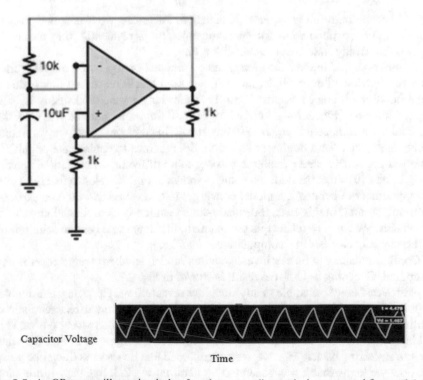

Fig. 2.5 An OPamp oscillator circuit that functions according to the integrate and fire model

In this circuit, the capacitor performs the integration operation. The waveform insert is the computed capacitor voltage for this circuit by means of circuit simulation.

2.7   Nonlinear Dynamic Properties

Earlier in this chapter, the population dynamics model was introduced. The limited population model then followed from this in the form of Eq. 2.3. The numerical form resulted in Eq. 2.4, called the logistic equation. Then, we used Eq. 2.4 to find a solution for the model as found in Fig. 2.2 for a given set of parameters and initial population P(0). This result verified that the limited population model worked as expected. That is, for any initial condition P(0), the population converges to a steady-state population. However, since this is a nonlinear equation, we are not done.

Here is the issue. We have shown the results for one set of parameters. Other parameters may yield results that have yet to be explored. The initial set of parameters converged to one solution. That is, the model reached a steady-state solution. Referring back to Fig. 2.2, it can be seen that the model reaches its final value quickly. In fact, convergence occurred after just three iterations. For a continuous time equation, steady state is defined as when $\dfrac{dP}{dt} = 0$. So, setting the model Eq. 2.3 equal to zero. The result is a quadratic equation. Therefore, we expect two possible solutions for the population for some parameters. Equivalently, this means for Eq. 2.4 that steady state occurs when $P_{i+1} = P_i$.

Another way of finding the steady state computationally is to compute "for a long time period." For example, since it was seen that P reached a final value in three iterations, it can be assumed that five times longer should be steady state. In our calculation of Fig. 2.2, we iterated for over 10 times this amount. It is clear that the final point computed is equal to steady state. This is a convenient way of finding steady-state value for a nonlinear equation. So, continue to explore the population equation for further steady states by using some different parameters. Choosing $a = 3.1$ and $P(0) = 0.6$, the iteration result is shown in Fig. 2.6. Notice that now, after a large number of iterates, the model converges to two values as would be expected for a polynomial. In this case, the steady-state solution is to oscillate between the two values. We can now repeat this procedure to find more steady-state solutions for different parameter sets in a computational manner.

Continuing along to further investigate this model, another parameter set is now computed. Choosing $a = 3.5$, the result is shown in Fig. 2.7.

Notice that now four stable steady states are revealed, and frequency has doubled in comparison with Fig. 2.6. Frequency doubling is a common characteristic of nonlinear dynamic systems. If we continue this process of increasing the value of "a" and then iterating the model, a result is reached that appears in Fig. 2.8 when $a = 4.0$. Referring back to Fig. 1.4, we had assumed that it was something like noise. But now we know that it was generated by iterating Eq. 2.3. In Chap. 1, this same

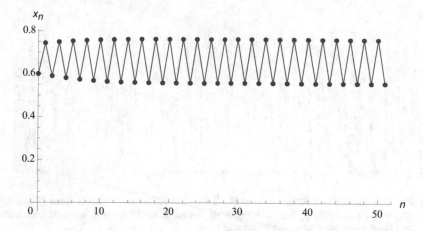

Fig. 2.6 N = 50 iterations of Eq. 2.4 for $a = 3.1$. X_n corresponds with population P, and n corresponds with the number of time steps (Generated using Mathematica's Wolfram Demonstration Project, Wolfram.com)

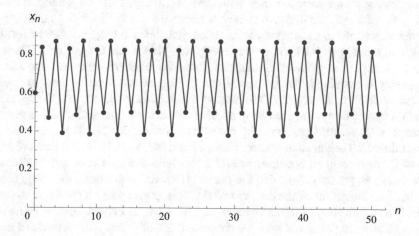

Fig. 2.7 Fifty iterations of Eq. 2.4 for $a = 3.5$. X_n corresponds with population P, and n corresponds with the number of iterations. (Generated using Mathematica's Wolfram Demonstration Project, Wolfram.com)

figure was introduced to you in Fig. 1.4 that might fool the researcher into believing their data was due to a random process. But now you know that it was generated by a real system and is deterministic and not random. This result is defined as Chaos. Looking more closely at Chaos in Fig. 2.8, it can be seen that a steady-state value was never achieved even though we iterated for $N = 50$ as in our previous examples. So, we see that the population equation does not have a final value for certain values of "a." Instead, every final value depends on the initial condition of $P(0)$. Another observation that may be seen from the Chaos result is that we might find that an infinite variety of frequencies have emerged. This process of investigating a

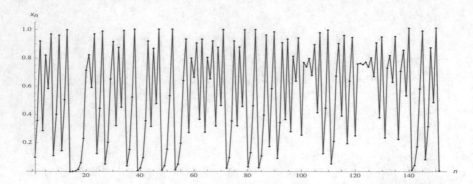

Fig. 2.8 One hundred and 50 iterations of Eq. 2.4 for $a = 4$. This is the Chaos result. Single variable obtained for variable x_n versus number of iterations n. (Generated using Mathematica's Wolfram Demonstration Project, Wolfram.com)

nonlinear system over a range of parameters is best summarized in a summary plot of parameter value versus the steady-state values. Notice that for $a < 3$ we find only a single steady state corresponding with the result in Fig. 2.2. Notice that just above the value of $a = 3$, we find two steady states as we did in Fig. 2.6. Continuing to higher values of the parameter, the iteration plot shows a large number of results. This corresponds with Chaos shown in Fig. 2.8. Notice that the graph of Chaos at first look appears to possess oscillations. But careful measurement reveals that no frequency is sustained. No frequency can be determined such as they are easily measured in Figs. 2.6 and 2.7. The lack of measurable frequency and the apparent randomness of Chaos are confusing at this point because the reader knows that all three time plots were generated by the same equation Eq. 2.4. Thus, they are all deterministic. The question arises: If I measured the data of Fig. 2.8 and I did not know it came from an equation, would I conclude that it is random? I think the answer is yes, possibly. Now that the reader is familiar with Chaos, they might look further into the details of the data to find that it is generated by a nonlinear system. That is a significantly different conclusion. This leads us into the next two chapters that will provide the reader with the characteristics of Chaos and methods of measuring and detecting it.

Problems

1. Iterate the Euler form of the limited population Eq. (2.4) for $a = 4$. Provide your answer as a plot of P version iteration number.
2. Repeat Problem 1 for $a = 3.5$. This is the chaotic condition.
3. Repeat Problem 2 for three different initial values of $P(0)$.
4. For Problems 1–3, find the number of iterations required to reach final value.
5. Solve the differential Eq. (2.3) for $l = 5$ and $a = 1$.

References

Crisp, K. (2019). Models for spiking neurons: Integrate-and-fire units and relaxation oscillators. *Journal of Undergraduate Neuroscience Education, 17*(2), E7–E12. Retrieved from https://www.ncbi.nlm.nih.gov/pubmed/31360134

Giordano, G., Blanchini, F., Bruno, R., Colaneri, P., Di Filippo, A., Di Matteo, A., & Colaneri, M. (2020). Modelling the COVID-19 epidemic and implementation of population-wide interventions in Italy. *Nature Medicine, 26*(6), 855–860. https://doi.org/10.1038/s41591-020-0883-7

Glass, L., & Mackey, M. C. (1988). *From clocks to chaos: The rhythms of life*. Princeton University Press.

Goldberger, A. L., & West, B. J. (1987). Applications of nonlinear dynamics to clinical cardiology. *Annals of the New York Academy of Sciences, 504*, 195–213. https://doi.org/10.1111/j.1749-6632.1987.tb48733.x

Bifurcation Mapping and Chaos

3

3.1 The Bifurcation Diagram Procedure

Now that the basic behavior of the logistic model has been observed, it is useful to formalize the approach that was used to organize the large amount of data that is produced when analyzing a nonlinear system. This approach is not only valid for the population model example that we used, but also for nonlinear dynamic systems in general. The end result is a bifurcation plot that summarizes the steady-state values for all initial conditions and parameters. A bifurcation diagram can be created by following the steps below:

1. Select an initial value of the system parameter/s.
2. Input an initial condition.
3. Generate the steady-state solution by iterating 100 time steps. Note, steady state may also be identified when the solution changes less than the resolution of the calculation.
4. Increment the initial condition and the initial parameter values. Repeat the solution for steady state with the new values of initial condition and parameter. All steady-state values are then plotted versus the system parameter for every initial condition.
5. This process is completed over the full range of initial conditions and system parameters. Implementing this procedure on the logistic equation population model in Chap. 2 yields the complete bifurcation diagram for the model as shown in Fig. 3.1.

Now study the details of the bifurcation diagram a bit further. It is clear that the value of $a = 3.25$ is a critical point on the diagram. Note that the graph is single valued below this value and dual valued above 3.25. This point is defined as a bifurcation point. Note that moving about the bifurcation point there are two possibilities. First, as "a" is reduced, the system remains stable at one solution. Second, as "a" is increased the system becomes unstable and two new solutions occur and the

G. Drzewiecki, *Fundamentals of Chaos and Fractals for Cardiology*,
https://doi.org/10.1007/978-3-030-88968-5_3

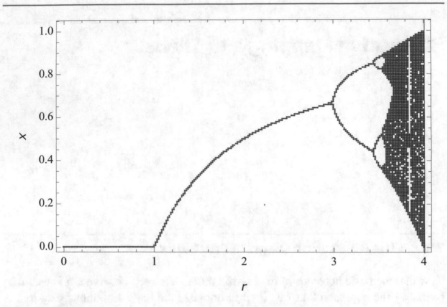

Fig. 3.1 The bifurcation plot for Eq. 2.4 the logistic equation. The rightmost side of the graph is Chaos. (Generated using Mathematica's Wolfram Demonstration Project, Wolfram.com)

model oscillates. At $a = 3.25$ point, it can be observed that the steady-state values abruptly change from 1 to 2 values and from no oscillation to oscillations. The process of jumping from one state into two is characteristic of nonlinear systems and is called period doubling. Thus, the bifurcation diagram provides very strong evidence of nonlinear system chaotic behavior and is a good tool to help identify a nonlinear system if the system is unknown. Notice also, moving toward higher values of "a," it can be seen that more bifurcation points occur at which period doubling happens again so that higher values of a result in greater and greater frequencies are generated until chaos is reached (Sander & Yorke, 2013). It was discovered that the ratio of the change between parameter values at adjacent bifurcation points is a constant (Feigenbaum, 1978). So that $(a_1-a_0)/(a_2-a_1) = 4.6692016$. The ratio is named after its discoverer, so it is defined as Feigenbaum's ratio. Now that we've seen the path to Chaos on the bifurcation graph, it is now time to define some of the attributes of a chaotic system. Recalling that the map was constructed over the full range of values for the initial conditions of the differential equation, it is impossible to equate any initial condition with a given steady state for the condition of Chaos. As Chaos has now been observed from the point of view of the iteration graph and the bifurcation plot, this leads to the more formal definition of Chaos. Following the definition of Sander and Yorke (2013), Chaos exists when:

1. For some parameters, all initial conditions result in a periodic dynamics
2. There is a sensitive dependence on initial conditions

Now review some of the other attributes of Chaos. As a result of points 1 and 2, the sensitive dependence of the initial condition, it can also be concluded from the bifurcation map and the long evaluation time of the steady state that a chaotic dynamic system is long-term unpredictable. Recall that all of the data points on the bifurcation graph are the long-term steady state values where we have computed to 50–100 times the initial transient time of the system. So, it can be seen that the rightmost values are many and varied such that it is not possible to find their values until after relatively long periods of time. Fortunately, studies of a chaotic system are not hopeless. Remember that the Chaos and bifurcation plot was produced by the simple population equation. Therefore, it should be expected that some kind of deterministic behavior must be observable and it is. But the observer must look over short time periods instead. That is, adjacent points in time must be governed by the differential equation of the model. To do this, introduce the Poincare plot, also named after the mathematician Poincare (Rasband, 2015). The Poincare plot is created by plotting the relationship between data points that are nearby in time. For example, in the population model, plot $P(t+dt)$ versus $P(t)$ for all value of time. This method is simplifying in that time becomes an implicit variable and the resulting graph only portrays the relationship between the system and its prior state. A Poincare plot for the conditions of Chaos is shown in Fig. 3.2. Notice that time does not appear on the plot. The time path of a chaotic population is shown proceeding from its initial condition of 0.6. The population values land on the parabola, which predicts the next value of population. A dashed line indicates the point of steady state.

The dashed line indicates when $X_N = X_{n+1}$ is the steady state. Following the vertical line beginning at the point 0.6 traces the time course of the solution for Chaos.

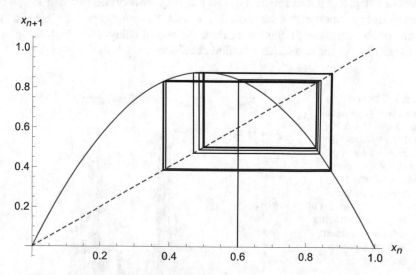

Fig. 3.2 The Poincare plot of chaos for the population model Eq. 2.4. X is the population value. (Generated using Mathematica's Wolfram Demonstration Project, Wolfram.com) where X_n is the current population, and X_{n+1} is the next population.

Another way of viewing Chaos is to observe its frequency content. This is obtained by performing the Fourier transform of a Chaos time series. This was performed on the Chaos generated by our population model by means of the fast Fourier transform (FFT) and further calculating the power at each frequency by squaring the magnitude values. The resulting power frequency spectrum is shown in Fig. 3.3. The power axis has been converted to a log scale so as to compress the values.

The results indicate that chaotic power has an inverse log-linear relationship with frequency. This means that Chaos possesses strong low-frequency content and loses power at high frequency, yet frequencies may extend to infinitely large values. Notice also that there exist some frequencies that are scattered throughout with large power. These appear as power "spikes" on the plot.

An inverse relationship with frequency is evident with increasing f. The long-term unpredictability of a chaotic system may also be viewed as an error that increases with time. For example, define the error of a solution as $E(t)$. In a chaotic system, the error increases with time so that $E(t) = E(0) * e^{ht}$, with $h > 0$. So, it is seen that error grows exponentially in a chaotic system. This observation leads to another useful method of testing for Chaos. For example, if there is a chaotic system of differential equations DE_1 and DE_2 and their solutions are $y_1(t)$ and $y_2(t)$, error for this single system is found as $E(t) = [y_1(t) - y_2(t)]$. The error will grow as $E(t = E(0) * 2^\lambda$ If it is found that $\lambda > 0$, where λ is the Lyapunov exponent. Then, the system is chaotic. Moreover, this provides an additional test of the existence of Chaos in such data (Shintani & Linton, 2004). Let us look at another example of the Poincare plot. This time apply the method to the integrate and fire oscillator illustrated in Chap. 2. That oscillator produced a very clear oscillation at a single frequency. But now look at this same oscillator with the Poincare plot. In this case, the output of the oscillator is plotted against the output delayed by 1 s. The plot is shown in Fig. 3.3 for two different initial conditions (Fig. 3.4).

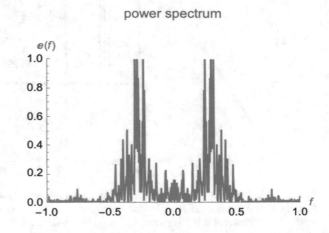

Fig. 3.3 The power spectrum of population chaos. Vertical axis is power magnitude. The frequency axis is normalized frequency f. Negative values of frequency are reflected about $f = 0$ as an artifact of Fourier. (Generated using Mathematica's Wolfram Demonstration Project, Wolfram.com)

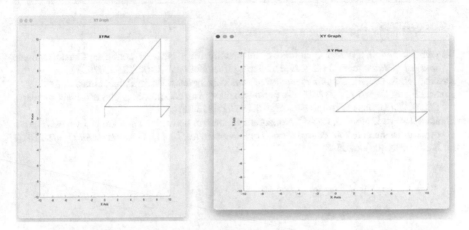

Fig. 3.4 The Poincare plot for the integrate–fire oscillator
The left panel is for an initial condition of 0, and the right for an initial condition of 5

First, the Poincare plot in this case has no recognizable geometry. Second, the plot returns to the same structure for any geometry. In the case where a system Poincare plot converges to a shape, that is not easily described. It is defined to be a "strange attractor." The shape of the attractor depends on the system, where some systems may possess simple geometries, such as a circle or ellipse. Moreover, since our oscillator always oscillates within the same attractor, it can also be defined as a limit cycle oscillator.

Problems

1. Using the iteration program from Chap. 2 Problem 3, plot the iterations for $a = 3$ and $a = 3.25$.
2. How many periods do you find for the parameters 3 and 3.25?
3. Using the bifurcation diagram of Fig. 3.1, find the value of "a" when the number of periods is 8.
4. If $a = 1$ and $L = 1$ in the population Eq. 2.1, find the two steady-state solutions for P.
5. As "a" increases from $a = 1$ to $a = 3.5$, list the number of periods that you see just above each bifurcation point.
6. Explain why you cannot see the number of periods for a greater than $a = 3.5$ on the bifurcation diagram.

References

Feigenbaum, M. J. (1978). Quantitative universality for a class of nonlinear transformations. *Journal of Statistical Physics, 19*(1), 25–52. https://doi.org/10.1007/BF01020332

Rasband, S. N. (2015). *Chaotic dynamics of nonlinear systems*. Dover Publications.

Sander, E., & Yorke, J. A. (2013). A period-doubling cascade precedes chaos for planar maps. *Chaos, 23*(3), 033113. https://doi.org/10.1063/1.4813600

Shintani, M., & Linton, O. (2004). Nonparametric neural network estimation of Lyapunov exponents and a direct test for chaos. *Journal of Econometrics, 120*(1), 1–33. https://doi.org/10.1016/S0304-4076(03)00205-7

Identifying Chaos

4

4.1 Chaos Versus Noise and Chaos in Experimental Data

As concluded in the last chapter, the identification of Chaos versus noise from data leads to quite divergent conclusions regarding the system under study. Simply put, the conclusion based on the observation of Chaos would be

$$\text{Chaos} = \text{deterministic system}, \text{versus } \text{noise} = \text{random system}$$

This chapter is meant for the researcher who is confronted with a data set that possesses variability of unknown origin. Fortunately, the methods of characterizing Chaos and nonlinear dynamic systems described in Chap. 3 are useful for identifying Chaos from noise in most situations. To begin, first review some of the methods of studying chaos.

4.2 Frequency Spectrum

One test might be to analyze the frequency content of the data by means of Fourier transformation. Chaos typically results in an inverse relationship between magnitude and frequency. This is illustrated in Fig. 4.1. This means that chaos is generally aperiodic. No dominant frequency peaks appear due to the bifurcation process and any given frequency will double.

Two features are noted. The low-frequency magnitudes are the largest. The high-frequency magnitudes become less with frequency, but extend out to relatively high frequencies. This is a consequence of frequency bifurcations. To differentiate the Chaos frequency spectrum from noise, consider a Gaussian random variation as was shown earlier in Chap. 1. The frequency spectrum for a Gaussian random variable time series is shown in Fig. 4.2.

The random noise frequency spectrum is broadband just like Chaos. The main difference is that the magnitudes of the spectrum are uniform as opposed to inverse

© The Author(s), under exclusive license to Springer Nature Switzerland AG 2021
G. Drzewiecki, *Fundamentals of Chaos and Fractals for Cardiology*,
https://doi.org/10.1007/978-3-030-88968-5_4

Fig. 4.1 An idealized
Fourier magnitude
spectrum of a chaotic time
series

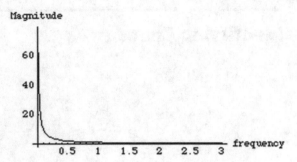

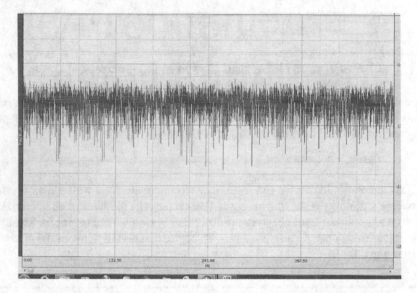

Fig. 4.2 Frequency spectrum of random noise. (Reproduced from Drzewiecki, 2015)

frequency. Hence, care must be taken to view a wide range of frequencies to clearly differentiate Chaos from noise. For example, if only the high frequencies are considered the two spectra are indistinguishable. It is only when the very low frequencies are examined do you see the strong inverse frequency effects appear. Therefore, limited availability of data may prevent the ability to recognize Chaos from frequency spectra. As such, the frequency spectrum may not be the best method to detect Chaos.

4.3   The Bifurcation Diagram

The bifurcation plot was introduced in Chap. 2 in association with the population model and is reproduced here in Fig. 4.3.

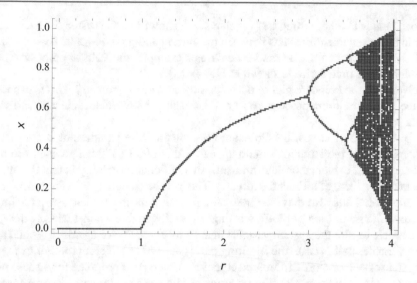

Fig. 4.3 The bifurcation plot of the logistic model. (Generated using Mathematica's Wolfram Demonstration Project, Wolfram.com)

The bifurcation plot provides a complete picture about a system's steady-state solutions for all parameter values and initial conditions. The key point here is the observation of bifurcation points where the system solution diverges into two solutions or simply where a single oscillatory frequency doubles its value. Referring to the bifurcation plot, it can be seen while progressing from left to right that more and more bifurcation points appear. So, in a chaotic system we see that bifurcations will progress to further and further bifurcations until the point of chaos is reached. The observation of progressing bifurcation points and a bifurcation plot is likely the strongest piece of evidence for a chaotic system that can be found (Giglio et al., 1981). But recall that this bifurcation plot for the population model example was generated in a computationally intensive manner by exploring all of the possible parameters and initial conditions. In some systems or experimental data, it may not be possible to examine a system's data so completely. Hence, a bifurcation plot may not be possible in which case the other less strong tests for chaos in this chapter may need to be employed. So, although a complete mapping may not be possible, the observation of period doubling provides a quick-test method for the presence of Chaos.

4.4 The Poincare and Phase Plots

Recall in Chap. 3 the Poincare plot was introduced as another method to characterize nonlinear dynamic and chaotic systems. It was applied to the population model for chaos parameters. The result in Fig. 3.2 clearly demonstrates a parabolic function. Other chaotic systems will also portray a clear geometry in their Poincare plots. This is another strong test for chaos as it shows that the system is deterministic in that a relationship exists between the current and previous data. By definition, this will not occur for true random data. For example, the Poincare plot of a pure random noise time series is shown in Fig. 4.4.

Note that the Poincare plot of noise has no structure or geometry. This is expected since a true random process has no relationship between the current and next data point.

In this case of chaos, the Poincare plot will indicate an attractor if the data is Chaos. Ideally, the attractor should appear as that of a Euclidean geometry to provide excellent evidence of determinism. An additional possibility is that the attractor takes the form of a Fractal geometry. This will be shown later in the Fractal part of this book. Note also that the phase plot is another means of identifying nonlinear system Chaos. Instead of plotting a single system variable against its time delay, it is possible to plot the system variable against its derivative or against a second system variable. In this case, the resulting plot is defined as a phase plot. For example, a system of two or more differential equations would be appropriate for a phase plot. Just as in the Poincare plot, the occurrence of a strange attractor or some type of geometric attractor would provide very good evidence that a chaotic data series is not random. Note that while strange attractors are a common structure that might be observed, the fractal structure is also possible in a phase plot. In the case of a fractal attractor, the attractor may look like an image.

In summary, the methods provided in this chapter offer a good start in identifying Chaos from random noise and will be useful more often than not. However, the reader should be aware that some nonlinear system chaos can be a challenge. To illustrate a situation when it is not possible to separate noise from Chaos, the

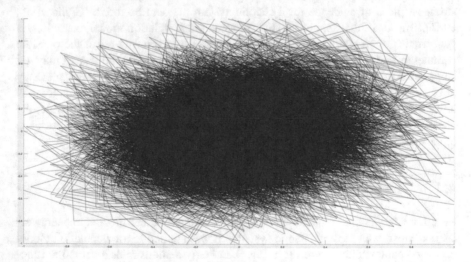

Fig. 4.4 The Poincare plot of a Gaussian random time series

model of nerve action potential firing of Glass & Mackey, 1988 provides such an example. The cell membrane action potential voltage is the primary indication of nerve cell activity. Nerve cell action potentials in the central nervous system would be expected to carry physiological information. Therefore, nerve cell activity might be expected to be deterministic, not random. A variable that may be used to quantify this nerve activity is the time between nerve action potentials or time between nerve events. Then for all the action potentials of the data recording the number of action potentials that do not occur in each time interval is summarized as a histogram in Fig. 4.5.

Looking at the data in Fig. 4.5 as a probability distribution, this is the probability distribution of a Poisson random process as shown in Fig. 4.6. Hence, common statistical methods may prove inadequate in detecting a deterministic process.

Fig. 4.5 Histogram of time intervals between nerve firings. (Reproduced from Deger et al., 2009)

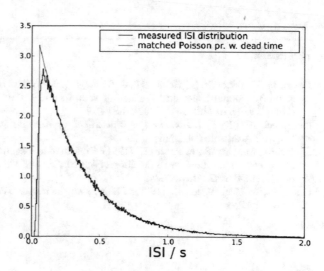

Fig. 4.6 Probability distribution of a random Poisson process

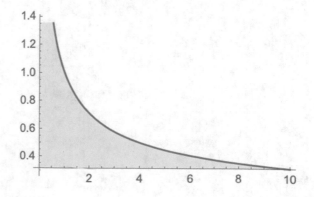

Problems

1. To verify that the population equation is chaos, create a Poincare plot by plotting the current versus prior point in an xy plot That is, plot $P(i)$ versus $P(i-1)$ using $a = 3.5$. Do you see an attractor?
2. To further verify chaos, plot $P(i)$ versus iteration number, i, for the population equation with $a = 3.5$ using an initial condition $P_1(0)$, of your choice, use the $a = 3.5$.
3. Repeat Problem 2. Use a different initial condition $P_2(0)$ and plot $p(1(i)$ together with $P2(i)$ on the same graph. You should see the two solutions for chaos diverging. This is "sensitivity to initial conditions.."
4. If your software has a Fourier transform function, apply it to your chaos iterations of Problems 2 or 3. If not, measure the iterations between peaks of your $p(i)$ plot for at least five peaks. Summarize your observations in terms of number of long and short periods that you find.

References

Deger, M., Cardanobile, S., Helias, M., & Rotter, S. (2009). The Poisson process with dead time captures important statistical features of neural activity. *BMC Neuroscience, 10*(1), P110. https://doi.org/10.1186/1471-2202-10-S1-P110

Drzewiecki, G. (2015). *Lab manual for biomedical engineering: Devices and systems* (2nd ed.). Cognella Academic Publishing.

Giglio, M., Musazzi, S., & Perini, U. (1981). Transition to Chaotic behavior via a reproducible sequence of period-doubling bifurcations. *Physical Review Letters, 47*(4), 243–246. https://doi.org/10.1103/PhysRevLett.47.243

Glass, L., & Mackey, M. C. (1988). *From clocks to chaos: The rhythms of life*. Princeton University Press.

Controlling Chaos

5

5.1 Introduction

A nonlinear system that generates Chaos consists of an infinity of solutions that can be stable or unstable. These can be seen on the rightmost side of a bifurcation plot such as shown earlier in Fig. 4.3. At first appearance, a chaotic system might seem to be useless to an engineer. However, the generation of Chaos possesses some intrinsic value to engineering and especially to biological design. Consider the possibilities when Chaos may be controlled.

Control of Chaos means that many system states are available for use. Thus, a variety of solutions are potentially available for a single chaotic system. This is a very efficient form of design. One system provides many solutions. For example, one biological structure may be employed by an organism to adapt to its needs and function. Consider the arterial system. Beginning with the aorta, the arterial system starts with a large blood vessel and manages to connect with vessels that are a thousand times smaller in diameter such that blood may be distributed throughout the various organ systems and tissues. This is accomplished by means of the branching process. Employing a very simple process creates a very complex structure. The repeated application of branching results in an infinity of small vessels that may be analogous to the concept of system states in space as opposed to time. In an engineering sense, this concept implies that one hardware design may include multiple functions. In order to fully implement a chaos system design, it is necessary to prevent the desired chaotic state from becoming unstable. This is referred to as the control of Chaos.

5.2 Method of Controlling Chaos

The control of chaos can be achieved via a few similar methods; the basic concept is only reviewed here. The Poincare map or phase plot was introduced in Chap. 3 as a means of studying nonlinear and chaotic systems. As an example, consider a

G. Drzewiecki, *Fundamentals of Chaos and Fractals for Cardiology*,
https://doi.org/10.1007/978-3-030-88968-5_5

sample phase plot in Fig. 5.1. The phase plot can be obtained in a variety of ways. For one plot, a system variable against a time-shifted version of the same variable, also referred to as the delay map. Additionally, it can be obtained by plotting a system variable against the derivative of that variable. Lastly, if the system has two state variables they may also be plotted against each other. The general idea is that time becomes an implicit variable so that only the nonlinear relationships of the system are displayed. Time then appears in the example phase plot of Fig. 5.1 as the arrow shows a counterclockwise motion around the attractor.

Two attractors of the many choices are shown. The arrow selects a point on this attractor that also coincides with a fixed observation time. This attractor was chosen from the many because it is a desirable function for this system. The right attractor is the desired attractor function in Fig. 5.1. Notice that the observation phase line now intersects two points. Since the two attractors appear quite close to each other on the phase plot, control can be achieved by forcing the system to move to the desired attractor. Control becomes a process of first testing the observation, and second, if not the correct attractor, then perturb the system in some way that forces it back to the desired attractor. This can be accomplished by various means as described by Boccaletti et al. (2000). In most cases when the attractors are adjacent to each other, only a small perturbance is required to correct the system to return to its desired attractor. By definition, the least energy needs to be expended if the perturbance is applied to the system initial condition. Alternatively, it may be practical to simply adjust a system parameter. Figure 5.1 shows a system that is producing only two trajectories on the phase plane. These trajectories are two solutions for the system. Each solution could have had different initial conditions or different system parameters. Mathematically, the process can be described for the solutions x_1 and x_2 by comparing them using subtraction, so that $e(t) =(x_1 - x_2)$. If x_1 is the required solution, then it is necessary to force $e(t) \to 0$. As time proceeds, the error $e(t)$ increases exponentially according to the Lyapunov exponent as $e^{\lambda t}$ where λ is the Lyapunov constant. Once the desired trajectory is achieved, changes are made in order to reduce the Lyapunov constant. This in turn causes the system to remain on the desired solution.

Fig. 5.1 Poincare plot of two illustrative attractors of a chaotic system. The dots and arrow indicate a single observation point on the attractor. Time proceeds in a counterclockwise path around this attractor. The axis corresponds to the system variables

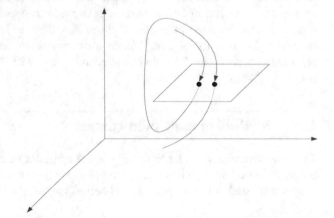

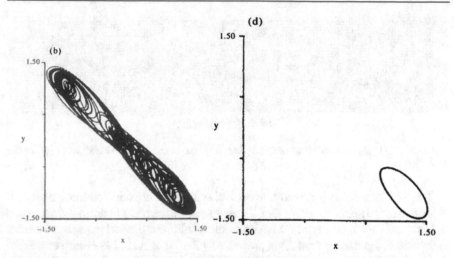

Fig. 5.2 Phase plot of voltage and current for a nonlinear RLC oscillator circuit. (Reproduced from Rajasekar et al., 1997). Phase plot (b) shows the circuit with chaotic oscillations and no control. Phase plot (d) shows a single period with control of chaos active

As another example, consider the Chua RLC diode electronic oscillator circuit that generates chaotic oscillations. The Chua oscillator consists of a series connection of resistor– inductance, capacitor, and parallel Chua diode (Chua, 1980). The diode provides the source of nonlinearity. The Chua circuit will be described further in Chap. 6 as a simple chaotic oscillator. The circuit oscillates as shown in Fig. 5.2 and demonstrates Chaos. The phase plot demonstrates a strange attractor in its phase plot. The researchers (Rajasekar et al., 1997) studied this circuit and applied perturbance methods that attempt to control the chaos into a desired oscillation. They showed that control of chaos permitted the circuit to oscillate at a single desired period as shown in Fig. 5.2d.

5.3   Control of Chaos Applied to Cardiac Arrhythmia Treatment

The cardiac electrophysiology pacemaker system is a nonlinear oscillator for which nonlinear systems methods lend themselves well to studies. In particular, the problem of cardiac arrhythmia may be a good application for nonlinear dynamics. The normal sinus arrhythmia resembles a chaotic oscillator. Abnormal rhythms such as tachycardiac and ventricular fibrillation are deadly and of great importance to study. Figure 5.3 provides the ECG for ventricular fibrillation, which eventually results in cardiac death. Figure 5.3 shows a loss of a stable oscillation of the heart and what has been proposed to be chaos (Dai & Schaeffer, 2010). The question that is relevant to this chapter is, can the control of Chaos methods presented here possibly restore an arrhythmia to normal?

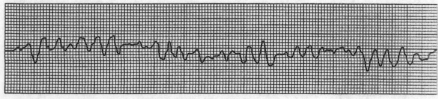

Ventricular fibrillation

Fig. 5.3 ECG of ventricular fibrillation that may be chaos. (Reproduced from Widmaier et al., 2023)

If it is assumed that the heart rate control system is a chaotic nonlinear system, it is reasonable to apply the control of Chaos to restore the heart to normal rhythm and out of the tachycardia rhythm. Medically this could be a lifesaving procedure since tachycardia is generally fatal if not prevented. (Garfinkel et al., 1992) have employed the control of Chaos to stop the condition of tachycardia. First, the Poincare plot is created for the patient from the data of beat-to-beat heart period. From this plot, the more normal physiological heart rate rhythms are identified in addition to the tachycardia rhythm. This leads to the idea of applying nonlinear dynamics to cardiology as a whole and is the subject of the next chapter.

Problems

1. Create a plot of the difference in the two initial conditions $(P_{10}(i) - P_{20}(i))$ versus i using your answers to homework Chap. 4, Problem 2.
2. You should see the solution difference plot increasing exponentially in Problem 1. This is chaos. Replot your difference graph of Problem 1 as a log plot. Is it now linear?
3. Attempt to minimize the difference plot by adjusting the values of the initial conditions or initial parameter value.
4. When you have successfully minimized the difference plot, show the iteration plot for those values. It should no longer appear chaotic.

References

Boccaletti, S., Grebogi, C., Lai, Y. C., Mancini, H., & Maza, D. (2000). The control of chaos: Theory and applications. *Physics Reports, 329*(3), 103–197. https://doi.org/10.1016/S0370-1573(99)00096-4

Chua, L. (1980). Dynamic nonlinear networks: State-of-the-art. *IEEE Transactions on Circuits and Systems, 27*(11), 1059–1087. https://doi.org/10.1109/TCS.1980.1084745

Dai, S., & Schaeffer, D. G. (2010). Chaos for cardiac arrhythmias through a one-dimensional modulation equation for alternans. *Chaos, 20*(2), 023131. https://doi.org/10.1063/1.3456058

Garfinkel, A., Ditto, W., & Weiss, J. (1992). Control of cardiac chaos. *Science, 257.*

Rajasekar, S., Murali, K., & Lakshmanan, M. (1997). Control of chaos by nonfeedback methods in a simple electronic circuit system and the FitzHugh-Nagumo equation. *Chaos, Solitons & Fractals, 8*(9), 1545–1558. https://doi.org/10.1016/S0960-0779(96)00154-3

Widmaier, E. P., Raff, H., & Vander, A. J. (2023).

Nonlinear Cardiology

6

6.1 Introduction: The Cardiac Arrhythmia

Nonlinear cardiology is the application of nonlinear dynamics theory to clinical cardiology (Goldberger & West, 1987). Perhaps the greatest area of application is the study of cardiac electrophysiology and arrhythmias. The cardiac pacemaker is likely the most well-known pacemaker in human physiology, and abnormalities of cardiac pacemaker cells provide the basis for clinical diagnosis and treatment of arrhythmias. In spite of the well-known origin of the heart rhythm, the dynamics of the heart rhythm is not yet clear. The study of cardiac arrhythmia remains one of the most challenging problems of cardiovascular research. The application of nonlinear dynamics to cardiology offers a different perspective on cardiac rhythm analysis and its functional understanding. So, we begin our analysis of the cardiac rhythm by treating the cardiac pacemaker in a nonlinear oscillator. In using this approach, the possibility of an infinite variety of cardiac rhythms is available for a single group of pacemaker cells depending on their parameters and interactions as we have seen in the population model (Chap. 2). The study of the cardiac rhythm has been approached on different anatomical levels as follows:

1. The cellular level: In this case, a single pacemaker cell may generate a rhythm on its own.
2. Membrane ion channel: At the member level, the regulation of ionic currents through the cell membrane is controlled by the ion channels of the cellular membrane. Researchers have used the Hodgkin–Huxley model (Hodgkin & Huxley, 1952) to test the possibility that arrhythmia might begin at the ion channel. Guckenheimer and Oliva (2002) used a computational Hodgkin–Huxley model to generate cell action potentials. They then did an exploration of the various ion channel parameters and initial conditions to locate instabilities. The model's attractor is shown in Fig. 6.1 for a solution that provided chaos. The results find a parameter set that provided unstable solution or Chaos. In summary, although it is possible that a single pacemaker cell is capable of causing arrhythmia on its

© The Author(s), under exclusive license to Springer Nature Switzerland AG 2021
G. Drzewiecki, *Fundamentals of Chaos and Fractals for Cardiology*,
https://doi.org/10.1007/978-3-030-88968-5_6

Fig. 6.1 Hodgkin–Huxley model phase plot during the chaos solution as computed by Guckenheimer and Oliva (2002). The cell membrane voltage, v, is plotted against the ion channel permeability, h

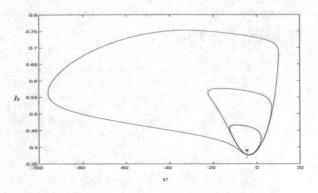

Fig. 6.2 System of two interacting pacemaker cells: the SA and AV nodes with interactive coupling that ultimately stimulate the cardiac ventricle

own, a narrow parameter set makes it an unlikely occurrence. This leads us to studies of cell-to-cell interaction as another possible source of arrhythmia.

3. Cell-to-Cell Interactions: Goldberger and West (1987) have performed studies of a two-pacemaker cell-to-cell interaction models. They begin with a simple nonlinear oscillator.

Consider the case of two cardiac cells interacting with each other. Referring to Fig. 6.2, the two cells are shown as group AV and group SA nodes. Common medical school physiology teaches that the cell group SA stimulates the group AV, and this group further stimulates the ventricular cells to electrically polarize and further lead to a mechanical contraction and the ventricular myocardium. Typically, no interaction between the two nodes is considered. A nonlinear oscillator was chosen by these researchers such that the SA and AV nodes may be allowed to interact. This is a good point to introduce some nonlinear oscillators.

6.2   Nonlinear Oscillators

At this time, it is useful to introduce a few oscillators that are nonlinear to study problems such as this. Goldberger and West chose the Chua oscillator circuit model (Chua, 1980) for this problem. The Chua oscillator is shown in Fig. 6.3. Figure 6.3 shows two oscillators: one to model the SA node and the second to model the AV node.

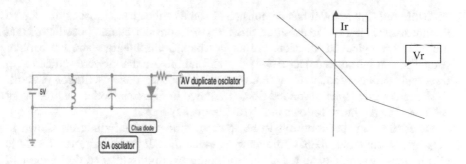

Fig. 6.3 The Chua oscillator circuit on the left to represent the SA node. A copy of the Chua is connected on the right to represent the AV node. The right graph is the Chua diode function curve. This oscillator contains a minimum of electrical elements. They are the inductor, capacitor, and diode

Fig. 6.4 Chua diode
voltage–current property

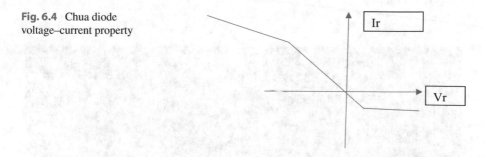

The Chua diode is special as seen from its voltage–current relationship in Fig. 6.4. It can be seen that the voltage–current law is that of a "negative resistance." Hence, it is a theoretical diode. But we can construct such a diode using additional circuitry or to use a tunnel diode that possesses some negative resistance. The Chua oscillator is unique in that it will generate chaotic oscillations as well as simple periods.

Returning to the problem of arrhythmias, this model of the SA and AV has been shown to produce a variety of arrhythmia that cardiologists commonly observe as demonstrated by Goldberger & West, 1987. The resistor connected between the SA and AV oscillators allows the coupling between the two oscillators to vary as might happen with a disease condition. So, the value of R simulates disease. Consider the condition when R=infinity. This is open circuit. In this case, the SA and AV oscillators cannot interact. Each pacemaker then operates at its own separate frequency. It is normal physiology for the SA to have a higher frequency than the AV pacemaker since the SA frequency is regulated. Now consider when the value of coupling resistance is reduced. This means that the SA and AV oscillators interact. At some value of resistance, the AV oscillator will adopt the SA frequency. So then, both oscillators are at the same frequency. Cardiologists call this normal sinus rhythm. Engineers define this as phase lock or capture. The AV node follows the SA oscillator.

While the Chua model has demonstrated some of the characteristics of coupled nonlinear oscillators, it is useful at this time to reconsider coupled oscillators that are more biological in function. Earlier in Chap. 2 the integrate and fire cellular oscillator was introduced. Here, two integrate and fire oscillators were coupled as in the Chua oscillator model. Coupling was varied to observe the oscillation patterns. A strong coupling was examined first and shown in Fig. 6.5. Notice that for every SA pulse the AV pulse follows in a 1-to-1 frequency lock.

After altering the coupling to be less, we find a mixed situation shown in Fig. 6.6. Notice that the SA frequency is twice the AV frequency, yet the AV is still in phase to some extent. This is a pathological rhythm that cardiologists call the AV block condition. As there are more model parameter sets that can be input to this model, it is clear that a rich set of frequencies are available from this dual nonlinear oscillator model, including chaos. In summary, a nonlinear oscillator model may simulate many of the arrhythmias observed by cardiologists in a very concise manner.

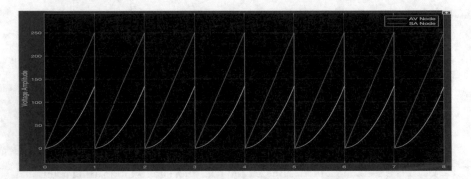

Fig. 6.5 Sinus rhythm shown as a phase lock of two coupled integrate and fire oscillators

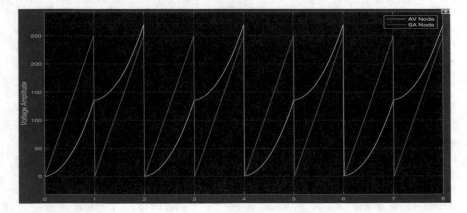

Fig. 6.6 Example of 2:1 rhythm or AV block of two weakly coupled integrate and fire oscillators

Continuing with the subject of nonlinear oscillations, a few more nonlinear oscillators are introduced. First, recall the resonant system from basic physics. The second-order differential equation for resonance is Eq. 6.1:

$$\frac{dy^2}{dx} = -\omega^2 * y$$

(6.1)

Equation 6.1 equation of resonance

where y is the oscillating variable and ω is the angular frequency of oscillation. A nonlinear term would need to be added to result in chaos. In addition, resonant systems would typically include an energy loss element. An additional forcing energy function is necessary to sustain oscillations. This kind of model can be viewed as a nonlinear resonant system that interacts with an external oscillation. An improvement on the resonant nonlinear system would be to add self-sustained oscillations. This can be found in the very concise Van der Pol nonlinear oscillator (van der Pol, 1927). The Van der Pol oscillator is described by the following nonlinear differential equation (Eq. 6.2):

$$\frac{d^2v}{dt^2} + \varepsilon\left(1 - v^2\right)\frac{dv}{dt} + v = \alpha\cos\left(\omega t\right)$$

(6.2)

The oscillating variable is v. The left side of this equation shows the terms for a resonant system with nonlinearity as indicated by the square term. The voltage in the square term is minus, indicating the positive feedback of voltage in the model. The middle term then serves two functions – to provide feedback and nonlinearity. The feedback of voltage provides a self-sustaining oscillation of voltage as required for any biological pacemaker cell. The right side of this equation is a forcing function , which in this case is a sinusoid. The parameter "ϵ" determines the amount of feedback and nonlinearity. The parameter "μ" sets the amount of external sinusoidal drive. Note that if the drive term is set to zero, this system will still sustain self-oscillation independently. A sample of Van der Pol oscillations is provided in Fig. 6.7 for three different values of μ.

Note that for very low external drive of $\mu=0.2$ the oscillations self-sustain and the oscillator functions independently. The right side of the figure shows the phase plot for the given parameter. The result is very clearly a circle shape. So, the phase plot attractor is a circle. For higher level of external drive, the attractors become more complex due to nonlinearity and are no longer a simple circle. The Van der Pol oscillator will exhibit bifurcations for some parameters as well as chaos. The rich set of dynamics generated by this simple system was used by Van der Pol himself to study arrhythmia in the heart.

In summary, this chapter has introduced and reviewed some nonlinear oscillators. Some of these oscillators were good models of the heart's pacemaker rhythm system. In particular, some oscillators were able to reproduce various cardiac arrhythmias. A single cell or oscillator was incapable of generating chaos, but coupled nonlinear oscillators provided a rich set of unstable oscillations similar to arrhythmia.

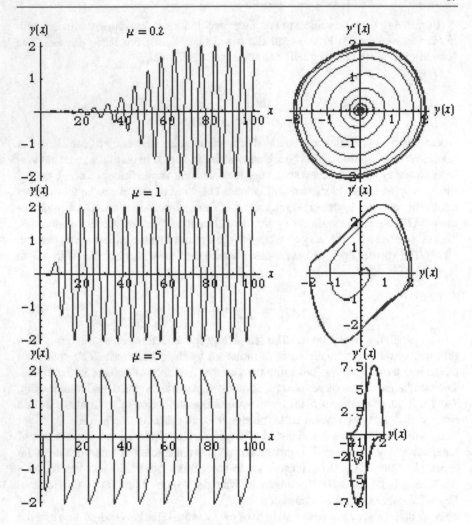

Fig. 6.7 Oscillations generated by the Van der Pol model equation (Eq. 6.2) for three different values of driving function μ. Left plots are the variable oscillations versus time. The right side shows the corresponding phase plots. (Plots were generated using the Mathematica's Wolfram Demonstration Project, Wolfram.com)

Problems

1. What are the three electrical elements required to cause chaos in an electrical oscillator?
2. If the SA and AV pacemakers are blocked, what rhythm does the ventricle take?
3. What property of the Chua diode sustains oscillations?
4. Are the arrhythmias seen in the SA and AV node pacemakers due to bifurcation periods?

References

Chua, L. (1980). Dynamic nonlinear networks: State-of-the-art. *IEEE Transactions on Circuits and Systems, 27*(11), 1059–1087. https://doi.org/10.1109/TCS.1980.1084745

Goldberger, A. L., & West, B. J. (1987). Fractals in physiology and medicine. *Yale Journal of Biology and Medicine, 60*(5), 421–435. Retrieved from https://www.ncbi.nlm.nih.gov/pubmed/3424875

Guckenheimer, J., & Oliva, R. A. (2002). Chaos in the Hodgkin–Huxley Model. *SIAM Journal on Applied Dynamical Systems, 1*(1), 105–114. https://doi.org/10.1137/s1111111101394040

Hodgkin, A. L., & Huxley, A. F. (1952). A quantitative description of membrane current and its application to conduction and excitation in nerve. *The Journal of Physiology, 117*(4), 500–544. https://doi.org/10.1113/jphysiol.1952.sp004764

van der Pol, B. (1927). VII. Forced oscillations in a circuit with non-linear resistance. (Reception with reactive triode). *The London, Edinburgh, and Dublin Philosophical Magazine and Journal of Science, 3*(13), 65–80. https://doi.org/10.1080/14786440108564176

Part II
Fractals

Fractal Geometry

7

7.1   Introduction to Fractals

Most scientists and engineers today are mostly familiar with Euclidean geometry, which is primarily the geometry of man-made objects that has been known for over 200 years. Examples are the circle, square, triangle, and so on. These geometries are described by algebraic formulae. Since these geometries are mostly man-made objects, they also possess size and scale. This chapter introduces you to another kind of geometry called the Fractal.

Fractal geometry is more consistent with natural or biological shapes and is therefore of greater interest to the bioengineer. The application of the Fractal also extends beyond biological structures to those shapes that have evolved in nature. Hence, the fractal has much greater application to nature than the classical Euclidean geometry. The mathematician (Mandelbrot & Mandelbrot, 1982) originally introduced the concept of the Fractal. Mandelbrot applied this new geometry to describe natural objects like the shape of a coastline, tree branches, branching plants, clouds, feathers, the surface irregularity, and mountains. Upon consideration of these objects, it is clear that Euclidean shapes such as straight line, circles, triangles, and the like would be very difficult to employ as models for these natural shapes. The key items to learn in this chapter are the formal definition of the Fractal and how to measure the Fractal object.

A Fractal possesses some properties that are quite different from Euclidean geometry. For example, the Fractal does not have a size or scale. It is constructed by a mathematical algorithm and must be constructed using iterative methods. Lastly, there are some Fractals that are purely mathematical and have no particular relationship with nature. You have already seen in the nonlinear dynamics chapters that dynamic systems possess a phase plot that may appear as a strange attractor. It is also possible that an attractor is a Fractal. If we can identify an attractor with Fractal properties, then it is not random since Fractals are deterministic. Therefore, Fractals can serve as another useful tool in identifying a chaotic versus random systems.

© The Author(s), under exclusive license to Springer Nature Switzerland AG 2021
G. Drzewiecki, *Fundamentals of Chaos and Fractals for Cardiology*,
https://doi.org/10.1007/978-3-030-88968-5_7

In this chapter, the Fractal will be defined, then it will be employed to model for some natural objects. Then, the fractal will be employed as a measuring tool that is more generally applicable to natural shapes and biology than classical Euclidean measurement.

7.2   Biological Branching Structure Examples

Since biology often uses branching for construction, it is a Fractal geometry as identified by Mandelbrot (Mandelbrot & Mandelbrot, 1982). For example, the pulmonary airway system of the human lung shown in Fig. 7.1 is a Fractal object. The repetition of the branching structure can be easily observed in Fig. 7.1.

Starting with the main airway, the trachea, the trachea branches into the two left and right pulmonary airways. At the end of the first branch of airways, the branching occurs again. This kind of branching is defined as dichotomous branching. Dichotomous branching then continues until the pulmonary alveoli or air sacs are reached. Notice that this is a very simple and organized process where one airway eventually branches into 1 million air sacs. Biology only needs to know the mechanism of how to branch, then it can construct immensely complex structures such as the entire lung.

Other structures within human physiology have been identified to be fractal in structure such as the cardiovascular arterial system. A possible microvascular bed dichotomous branching scheme is shown in Fig. 7.2.

Goldberger and West (1987) have recognized other examples of the fractal physiological structures, for example, the branching of neurons and the tubule system of

Fig. 7.1 Fractal branching structure of the human pulmonary airway system. (Mandelbrot & Mandelbrot, 1982)

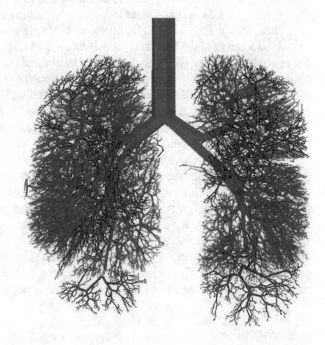

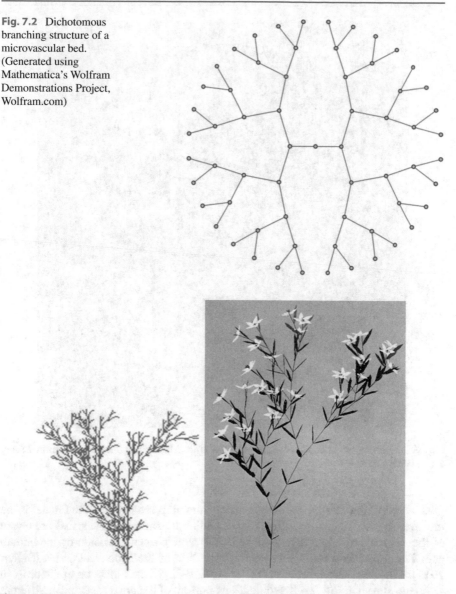

Fig. 7.2 Dichotomous branching structure of a microvascular bed. (Generated using Mathematica's Wolfram Demonstrations Project, Wolfram.com)

Fig. 7.3 Fractal plant structures generated using the L-system algorithm. (Reproduced from Prusinkiewicz et al., 2012)

the liver. It is easy to locate fractal branching in our environment from the shape of most plants and trees. The same branching scheme was implemented by means of a computer graphics algorithm defined as the L-system (Prusinkiewicz et al., 2013). It was found that many realistic plant shapes could be modeled (Prusinkiewicz et al., 2012) using an L-system. A sample plant produced by the L-system algorithm is shown in Fig. 7.3. The many plant structures produced by this graphics-recursive algorithm are quite realistic to the human eye.

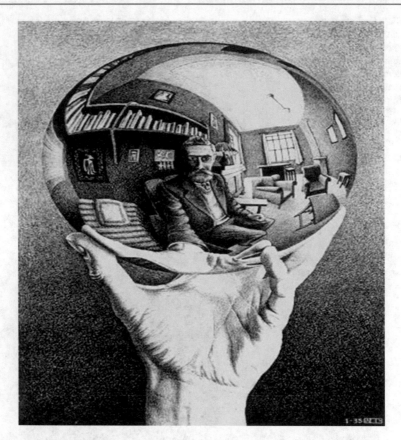

Fig. 7.4 Self-portrait at repeated scales by the artist M. Escher. (Reproduced from Escher et al., 1982)

The use of Fractals in artworks is also common. It may be that the Fractal is the most representative shape to draw nature with. For example, most artists are aware of the 1/3 scale rule that is thought to be the most pleasing division by the human eye. The 1/3 ratio also appears as the first ratio of the Fibonacci series (Sigler, 2012). The works of M. Escher (Escher et al., 1982) of the reflection of the artist in a sphere, shown in Fig. 7.4 illustrate a self-portrait of the artist repeated at different scales. Other works by Escher provide similar examples of the use of Fractals in art.

7.3 Fractal Time and the Mathematical Fractal

The concept of Fractal geometry may also be extended to time signals. In this case, the time waveform is viewed as geometry where the signal versus time dimensions are replaced by x versus y dimensions. This interpretation of a time signal may be considered an additional kind of biomedical signal processing. In this chapter, some examples of applying the Fractal to biomedical research will be provided. It will be seen that Fractal time can be related to the concepts introduced earlier in the chapter on nonlinear dynamics and the strange attractor where the Fractal time is simply a type of strange attractor of the dynamics of a system. By themselves, Fractals can stand alone as an area of mathematical science that will be referred to here as the mathematical or deterministic Fractal. This kind of Fractal may possess no direct application to biology but furthers our understanding of the Fractal as a mathematical concept. The deterministic Fractal can fool the human eye into looking like a real natural object, though it is not.

7.4   Fractals Defined

The Fractal is most simply defined as a geometry that possesses self-similarity in that it is independent of scale or scaling operations. A fractal therefore possesses no size since the object will appear identical at any scale. Once the scaling properties of a Fractal image are known, it offers the possibility of very dramatic reductions in data storage requirements where image data reduction of as much as 100 times has been possible (Barnsley & Hurd, 1993).

   To learn the fundamental definition of the Fractal, a Fractal is now generated by example. Begin by starting with a straight line as in Fig. 7.5. For ease, assume the length of the line is $L = 1.0$ and it has been divided into thirds denoted by each point. The division by 1/3 is then the scale factor that has been applied to the line. Let the scaling factor be $r = 1/3$. So, the line now consists of $N = 3$ pieces. Repeat the division again this time with $r = 1/4$ and reassemble the pieces differently, as shown in Fig. 7.6, but keeping the length $L = 1$ for the object. Figure 7.6 shows three

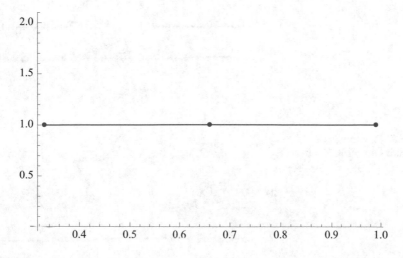

Fig. 7.5   Straight line with 1/3 divisions indicated at each point and the total length $L = 1$

iterations of this process where the middle line segment is replaced with a triangle segment as in iteration n = 1.

Next, dividing the line by 3 so that now the line has been divided into pieces of length 1/3 creates the fractal object. Define the length of each piece as the quantity "r," so that $r = 1/3$. Now reassemble the pieces of r into the original length L The reassembly here is shown in Fig. 7.6. Note that other assemblies could have been made and will be shown later. But follow the fractal of Fig. 7.6.

It is important to note that the original line now contains four of the pieces $r = 1/4$. So, we then define the number of pieces r as $N = 4$. This process of division can be continued indefinitely. For one example further if we divide by 8, or let $r = 1/8$, we now have $N = 8$ pieces. These are all reassembled into the original line $L = 1$. The structure for $N = 8$ is shown in Fig.7.7.

The object constructed by the above procedure is a very simple fractal called the von Koch snowflake (Koch, 1904). This simple construct may now be used to define the Fractal dimension. Continuing on to $N = 10$ iteration reveals the Fractal called the von Koch snowflake. Other iterations are shown in Figs. 7.8 and 7.9.

Note the natural resemblance to a real snowflake. Also note that if you magnify any portion of the snowflake, it always resembles the basic object that we started with in Fig. 7.6.

Fig. 7.6 Reassembly of line divisions into original line

Fig. 7.7 Reassembly of structure for $r = 1/8$, $N = 8$

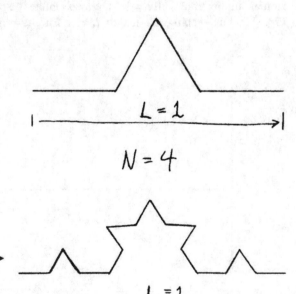

Fig. 7.8   One, two, and three iterations of the object in Fig. 7.6

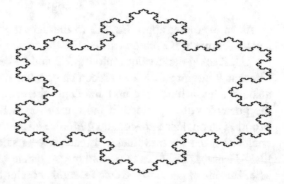

Fig. 7.9   Ten iterations of
the object in Fig. 7.6 yield
the Fractal snowflake
shown here. (Generated
using Mathematica's
Wolfram Demonstration
Project,Wolfram.com)

7.5   Defining the Fractal Dimension

In the derivation above, the value r = scale factor or, alternatively, after division
r = the length of a piece of the original line. Note also that after each division the
number of pieces, N, is related to the scale factor according to $r = 1/N$ that may also
be written as

$$N = r^{-1} = 1 = L \quad (7.1)$$

Since the snowflake fractal begins from a line object, Eq. 7.1 has an exponent of
1. Repeating the same example using an area object (e.g., a square) that is divided,
it would result instead in an exponent of 2. Repeating the exercise with a volume or
three-dimensional object, the exponent would be 3. For example, a cube is a three-
dimensional object. So for any dimension object, Eq. 7.1 may be generalized by
allowing the exponent of r to be an integer number D so that the general Fractal
relationship becomes Eq. 7.2, where D = the fractal dimension:

$$N = r^{D} \quad (7.2)$$

Note that the value of D in each example presented so far reveals the dimension-
ality of the Euclidian object that was analyzed. Hence, the fractal exponent provides
a conceptual measurement of the spatial dimension of the object. Measuring D of a
geometry therefore can be thought of as measurement of "space filling." In our
above snowflake example, the value of D indicates how close the geometry
approaches a true volume. While this kind of reasoning is only an aid to our understanding of the geometric characteristic of the fractal, it is best to know the D is
more accurately a measurement of the scaling property of the geometry.

Since it is often desirable to measure the Fractal dimension, the general Fractal law may be rearranged into a more convenient form to calculate D from r and N that are the measured quantities. Taking the log of Eq. 7.2 yields the log form of the fractal dimension:

$$-\frac{\log N}{\log r} = D \tag{7.3}$$

As an example, apply Eq. 7.3 to measure the Fractal dimension of the fractal snowflake that was created earlier. For the fractal of Fig. 7.6, recall that $N = 4$ and $r = 1/3$. Placing these values into Eq. 7.3 and the result is $D = 1.25$. Notice that D is less than 2 but greater than 1. It can be concluded that the object is not an area and also not a line. Hence, we find that D provides some interpretive information about the geometry of an object. If the dimension is higher and approaches $D = 2$, it becomes an enclosed space, and therefore an area when $D = 2$ but yet is not yet an enclosed space for less than 2. It can only be said that the Fractal is more "area-like." Hence, D provides more of a measurement of shape and scale rather than size as in Euclidean geometry. Some researchers refer to the value of D as a measure of "space fillingnes." In either case, D is a descriptive value of an object. We turn next to measuring the Fractal dimension of an object.

7.6   Measuring the Fractal Dimension

Now that the fractal has been defined, it can be applied to measure the Fractal dimension of a Fractal object. It is also possible to apply the Fractal concept to real-world structures and to time-series data. Time data may be translated into spatial data. In transforming time data into spatial data, the Fractal provides another means to study the nonlinear dynamic systems that generate chaotic time series. Hence, fractal dimension may be included as another test of chaos and, more importantly, that a given data set is deterministic as opposed to random (Katz, 1988). Thus, Fractals have become a useful tool in biological signal processing as a test of determinism for time-series data. A deterministic system will generate Fractal data and possess a measurable dimension. A few methods of measuring D will now be presented. The first method stems directly from the defining equation for D in Eq. 7.3. For example, if we can measure r and N, these values are inserted into Eq. 7.3 to find D. This method was first invented by Mandelbrot (Mandelbrot, 1967). He defined the Fractal as a means of correcting for map scale in the measurement of length. Mandelbrot discovered that when measuring the length of a coastline on a map he would find different lengths depending on the ruler size. This led to his development of a solution for the problem that turned out to be the Fractal.

Consider the sample coastline map in Fig. 7.10. In the simplest case, a ruler is applied to the coastline in a sequential end-to-end measurement until the end is reached. The number of rulers needed to cover the entire coast is N, so the length L is $L = N*r$. While this is seemly a straightforward measurement, a complication arises if the measurement is repeated for a different map scale. For example, if the

Fig. 7.10 Sample
coastline of Great Britain
length as measured by
sample length r.
(Reproduced from the
Fractal Foundation with
permission from the
fractalfoundation.org)

N=19
r=2

map scale is increased, r is effectively smaller. It is easy to realize that L will now increase as r is reduced. Of course, it is the same coastline so that the length must be constant and the error is a result of the ruler size, r, chosen. It is clear that the coastline measurement changes with scale. This is a consequence of its Fractal property. Hence, length has no utility. Instead, measure the coastline's Fractal dimension. To measure D, record the number N of r lengths that are needed to cover the coastline. Then repeat the measure for different values of r to collect a data set of N and r values. Using the coastline data of $N(r)$ points, construct a log–log graph of log N versus log r as shown in Fig. 7.11.

The plot of Fig. 7.11 reveals a linear relationship between $\log(N)$ and $\log(r)$, which is expected for a Fractal according to Eq. 7.3. Also, by Eq. 7.3 it is found that the slope of this line is the value of $D = 1.25$ as reported by Mandelbrot (Mandelbrot & Mandelbrot, 1982). Each point on the graph represents a separate measurement trial. The data may be analyzed in a statistical manner by performing a linear regression fit to the data to find the slope of the line in Fig. 7.11, which by Eq.7.3 is the value of D. Consider the case when $D = 1$. It is then found that $N = 1/r$ or that the number of points is inversely proportionate to the measurement length. It is interesting to note that the dimension of the von Koch snowflake was 1.26. This means that the coastline of Great Britain and the snowflake has a very similar fractal property.

Hence, as the measurement size r becomes smaller, the number of data points increases. Since by simple linear measurement the coastline is $N*r = L$, it can be seen that length will depend on r as was discovered earlier. Alternatively, D is not subject to this problem and is only representative of the Fractal geometry of the coastline.

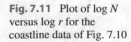

Fig. 7.11 Plot of log N versus log r for the coastline data of Fig. 7.10

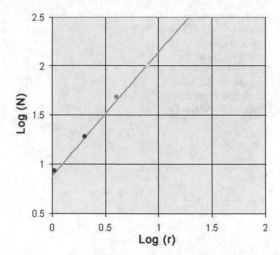

Hence, the value of Fractal geometry is clear in the case of this natural structure and standard Euclidean geometry yields unreliable information since it does not model the object realistically.

The line ruler, r, measurement method provided above was useful for illustrating the concept of Fractal D measurement but is not the most easily implemented or most accurate. For example, try to use this method on the fractal snowflake presented earlier and the task becomes impractical. A more practical and accurate method for measuring D is the box counting method (Li et al., 2009). The box method is implemented by placing a grid of equal-sized square boxes over the fractal object and counting the boxes that cover the object. As before, the number of boxes is counted that possess the Fractal to find N and the box size is now r. Then, Eq. 7.3 is used to find D as before. Again, a log–log plot is used to graph multiple trials and linear straight-line regression can be used to find the slope equal to D.

The box counting method is easily extended to other structures besides planar objects. For example, the concept is applied to volumes by using a fixed volume cube of size r^3. Then, a volume object is examined by counting the number of cubes that are required to enclose the entire object. The box volume, $r^3 = V$, is then changed and a new count is obtained. Several different box volume sizes are applied to the object to find the volume dimension D.

The fractal formula is slightly modified to reflect a volume situation as follows:

$$\log N = -D \log V \tag{7.4}$$

The volume equation requires that the box count and box volume are linear log–log for a fractal. Another form of Eq. 7.4 is to convert from box volume to box mass by multiplying by the mass density factor. This yields the mass dimension form of D:

$$\log N = -D \log M \tag{7.5}$$

The choice of using Eqs. 7.3, 7.4, or 7.5 to find D is only a matter of convenience and the form in which data is available.

7.7 Statistical Fractals

The preceding presentation of Fractal measurement was provided directly to the variables of interest. It is equally valid to apply the Fractal measures to the statistical values of the variables. Equation 7.3 is then modified to reflect a statistical measure such as mean value. For the mean values, the defining Fractal relation becomes $D_{mean} = \log N_{mean}/\log r$, where N is now the mean box count to find the statistical D_{mean}.

We may continue along with this idea by next applying the Fractal law to the standard deviation of N versus r. Further extending this idea, higher and higher order statistics may be applied to an object. This analysis can yield a different value of D for each statistic. This object is multi-fractal. In the case that it does not and it is found that every statistical D is equal, then the statistical fractal is defined to be a uniform Fractal.

Problems

1. List five fractal objects that you can see from your surroundings.
2. Identify four physiological systems that have a fractal geometry.
3. Use the ruler method to find the fractal dimension of the Koch snowflake in Fig. 7.9. It may be helpful to copy and enlarge the figure.
4. Will enlarging the Fig. 7.9 alter its fractal dimension D?
5. Find a tree outside and count the number of branches at each order up to branch order 5. For example, order 0 is the trunk, order 1 is the first branch, and order 2 is the second branch. Provide your results in a graph of number of branches versus order number. Can you observe a pattern?

References

Barnsley, M. F., & Hurd, L. P. (1993). *Fractal image compression*. AK Peters.

Escher, M. C., Bool, F., Locher, J. L., & Escher, M. C. (1982). *Escher: With a complete catalogue of the graphic works*. Thames and Hudson.

Goldberger, A. L., & West, B. J. (1987). Fractals in physiology and medicine. *Yale Journal of Biology and Medicine, 60*(5), 421–435. Retrieved from https://www.ncbi.nlm.nih.gov/pubmed/3424875

Katz, M. J. (1988). Fractals and the analysis of waveforms. *Computers in Biology and Medicine, 18*(3), 145–156. https://doi.org/10.1016/0010-4825(88)90041-8

Koch, H. (1904). *Sur une courbe continue sans tangente obtenue par une construction gÈomÈtrique ÈlÈmentaire*.

Li, J., Du, Q., & Sun, C. (2009). An improved box-counting method for image fractal dimension estimation. *Pattern Recognition, 42*(11), 2460–2469. https://doi.org/10.1016/j.patcog.2009.03.001

Mandelbrot, B. (1967). How long is the coast of britain? Statistical self-similarity and fractional dimension. *Science, 156*(3775), 636–638. https://doi.org/10.1126/science.156.3775.636

Mandelbrot, B. B., & Mandelbrot, B. B. (1982). *The fractal geometry of nature* (Vol. 1). WH freeman.

Prusinkiewicz, P., Hanan, J. S., Fracchia, F. D., Lindenmayer, A., Fowler, D. R., de Boer, M. J. M., & Mercer, L. (2012). *The algorithmic beauty of plants*. Springer.

Prusinkiewicz, P., Lindenmayer, A., Fracchia, F. D., Hanan, J., & Krithivasan, K. (2013). *Lindenmayer systems, fractals, and plants*. Springer.

Sigler, L. (2012). *Fibonacci's Liber Abaci: A translation into modern english of Leonardo Pisano's book of calculation*. Springer.

Fractal Time and Noise

8

Now that the Fractal has been defined, turn now to examine some of the different kinds of Fractals.

8.1   Three-Dimensional Fractals

Up until now, the fractals discussed have been two-dimensional objects. The fractal concept need not be restricted to 2D structures. There may also be three-dimensional fractals. These structures refer to volumes and may be analyzed as described in Chap. 7 by means of the volume box or mass dimension for a constant density object. For example, review some of the fractals examined so far. The fractal snowflake was found to possess $D = 1.26$. The US coastline has a similar value at $D = 1.25$. A perfectly flat surface is exactly an area with $D = 2$. Adding roughness to the surface increases the dimension to, for example, $D = 2.8$. Applying this idea to the surface of planet earth that may be considered a rough surface for which it has been determined that $D = 2.15$ for planet earth (Mandelbrot & Mandelbrot, 1982).

Other familiar objects that approach a 3D are clouds, the human head, and celestial galaxies (Mandelbrot, 1983). While the dimension value clearly provided some idea of the type of object that it is representing, inconsistencies in the object may lead to deviation from the fractal law. This problem can be handled by statistical fractals as described in Chap. 7.

Since most natural objects are not perfectly regular, they may be more easily expressed as an object that deviates from a pure mathematical relationship. This concept applies to the fractal as well such that the statistics of a geometry follows a fractal relationship as in Chap. 7.

G. Drzewiecki, *Fundamentals of Chaos and Fractals for Cardiology*,
https://doi.org/10.1007/978-3-030-88968-5_8

8.2   Fractal Time

A time-series waveform may at first consider to possess a self-similar geometry. Take, for example, the time-series data of Fig. 8.1.

In this example, the data was generated by a list of Gaussian random numbers such that it is just noise. But the similarity to a fractal geometry, such as the snowflake, is evident. Hence, it is reasonable that Fractal analysis may be applied to time-series data with the assumption that it is a self-similar geometry. Of course, this is not exactly correct since the one axis is actually time as opposed to a physical dimension. Outside of this exception, it is possible to apply fractal analysis to a time series. In general, the fractal concept may be applied in this case, provided that the relationship between the two different axes is a constant scale factor to transform the time axis. If a fractal dimension exists for a time series, it is then referred to as fractal noise. The value of D is determined as if the time series were a physical geometry. Some care must be taken though to ensure that the value of D is statistically significant by using the analysis of variance method as described by Katz (1988). Additional time series may be constructed from the Gaussian random series.

8.3   Fractal Time Generation

Other fractal time series may be generated from the random Gaussian time series.

For example, the random walk time series is derived from the random Gaussian by accumulating fixed random steps in a given time interval. The length of each step is assumed to be constant, but the direction of the step is then determined by the random Gaussian process. For example, if the step is of length 1, then the direction is either +/−1 and added to the origin. A random walk is shown in Fig. 8.2.

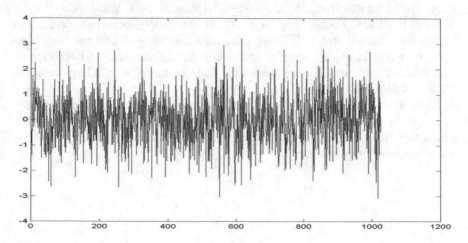

Fig. 8.1 Example of a Gaussian random time series (white Gaussian noise)

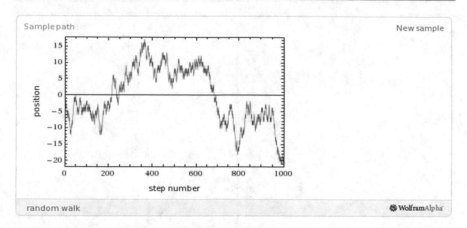

Fig. 8.2   Random walk time-series data. Also a fractal noise

Since noise is common in physical and biological data, the use of fractal time is a useful means of describing the characteristics of noise. It is thought that statistically significant D value suggests the presence of determinism. That is, a physiological signal such as the EEG contains information that the fractal dimension may detect (Rapp et al., 1989).

Most often noise is described by the standard Gaussian probability function. While the Gaussian is the most widely used model that is employed to deal with random noise in measurement and probability and statistics, the Gaussian normal probability function is given as $P(x) = e^{-x^2}$, where x is the random variable with a mean of 0 and standard deviation of 1. A normalized Gaussian probability distribution is shown in Fig. 8.3.

The Gaussian probability function can be used to generate a random time series. For example, a series of Gaussian numbers can be treated as a time series with a fixed time step Δt, resulting in a Gaussian normal time series as generated in Fig. 8.4.

Biomedical signal processing generally assumes that noise is of the form of Gaussian as in Fig. 8.3. This is a convenient assumption since the average of Fig. 8.4 Gaussian noise must by definition yield zero. Hence, the method of signal averaging relies on this basic property of the Gaussian noise. Unfortunately, the Gaussian random is rare in nature (Frank, 2009). One physical example of a Gaussian time series comes from radioactive decay. Other true Gaussian random time series are difficult to find. As you will see later in this chapter, most biological signals are fractal as opposed to Gaussian random. Alternatively, the random walk scaling properties are a better approximation of biological noise and variation. The random walk has been applied to model such phenomena as heart rate variability, cardiac blood flow variation, and river flow data (Goldberger et al., 2002; Hurst, 1951).

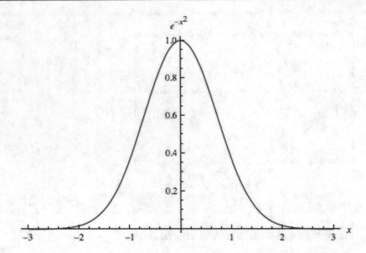

Fig. 8.3 The Gaussian normal probability function for a standard deviation of 1 and mean value of 0

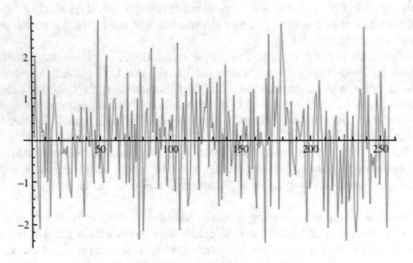

Fig. 8.4 Random noise time series of the probability function in Fig. 8.3

8.4   Fractal Time in the Frequency Domain

While the scaling properties of a Fractal series can be measured using the standard ruler method to find the fractal dimension D, the analysis of the time series by the Fourier methods is an alternative approach. Application of Fourier analysis may also be a convenient approach due to availability of computational software.

The frequency properties of Fractal time series are now examined. To begin, examine the random Gaussian time series of Fig. 8.3. A time to frequency transformation is performed by means of the fast Fourier transform (FFT) yielding magnitude versus frequency X(f). The magnitude data is then further transformed into power by computing $X^2(f)$. Since the range of power will normally be large, the log power is then computed. Hence, in the frequency domain, it is most common to relate log power versus frequency. The resulting power spectrum of Fig. 8.4 is shown in Fig. 8.5.

It can be observed that the log power spectrum extends over several orders of magnitude and the frequency range is very wide, extending from 0 to 1 kHz. This power spectrum gives the common name of the Gaussian random time as "white noise" due to the presence of a broad spectrum. Moreover, note that the linear fit indicates an average uniform power magnitude. This same analysis can be applied to any fractal time series. For example, the random walk is analyzed to reveal the power spectrum in Fig. 8.6.

In this case, it is found that the log power is related by the inverse square log frequency. In general, for any fractal time series, it can be found that $\log X = -\beta \log f + c$, where β is the slope of the log–log relationship that relates to the fractal dimension D. (Peitgen et al., 2012).

Figure 8.5 shows that the slope of the Gaussian random power spectrum was zero, so then $\beta = 0$ for Gaussian random time and the Gaussian noise is not fractal.

Figure 8.6 shows that $\beta = 2 =$ for the random walk time series. Hence, it has fractal properties.

In summary, the log power–log frequency relationship may be used to reveal the kind of fractal noise that may be present in a time series. In general, a slope of $-\beta$ may result, but a negative slope is characteristic of fractal time recalling the concept of chaos. We find also that a time series that is chaotic also results in a negative log power–log frequency slope (Peitgen et al., 2012). This idea was introduced in Chap. 4 where methods of testing for chaos were provided. It is reasonable to understand

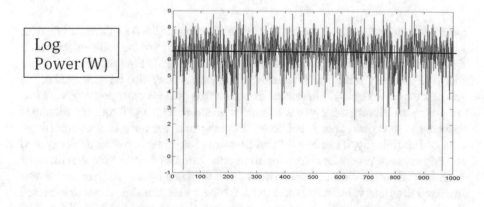

Fig. 8.5 Log power of Gaussian random time series in Fig. 8.4. Black line indicates a linear best fit to data

Fig. 8.6 Sample fractal time waveforms. Top wave is Gaussian random noise. The middle and bottom waves are frequency -adjusted noise waves. The left graphs are the corresponding frequency magnitude transfer spectrums that result from each noise waveform. (Reproduced from Peitgen et al., 1988)

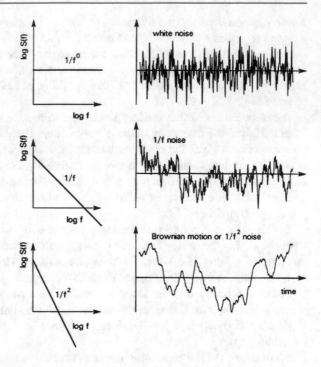

that a fractal is created by means of a recursive process, that is, since chaos can result from an iterative process, that chaos is also fractal in property.

To summarize fractal time in the frequency domain, fractal, in general, possesses an inverse magnitude–frequency relationship. Hence, it can be written as

$$X(f) \propto \frac{1}{f^{\beta}} \tag{8.1}$$

where β is the exponent of frequency. Notice that slope is $\beta = 0$ for Gaussian random noise as seen in Fig. 8.5 and $\beta = 2$ for the random walk. So, the value of β helps to determine the kind of fractal time series. Generally, fractal time follows Eq. 8.1 so that it is inversely related to frequency. Equation 8.1 may also be converted to log form as expressed above. Therefore, fractal signals possess most power at very low frequency and diminishing power at high frequencies. The low-frequency content is reasonable to expect since it is known that long time memory is characteristic of Chaos. Equivalently, it can be said that chaos has long time correlations as opposed to truly random processes. Now the time correlation of fractal time and chaos is examined. Another view of fractal time can be obtained by studying the signal during fixed time intervals. Due to the inverse frequency relationship, it can be expected that the range of the data will be greater for long time intervals and small for short time intervals. This relationship was discovered by Hurst (Peitgen et al., 2012) that leads to the Hurst dimension.

8.5 The Hurst Dimension

Another approach to analyzing Fractal time series derives from a basic observation of fractal time series. That is, Fractal time series exhibit an occasionally large change in value. If fractal time data is analyzed by means of its probability distribution, it is found that these large value changes appear with low frequency but are much greater than average. It is often referred to as large probability distribution "tails." This observation was first made by the river engineer Hurst (1951), who discovered that river flow data exhibits rare events of extreme large flow. But otherwise the flow data varies in a small range about the average flow. Brief time intervals of data extremes are characteristic of fractal time. At the time of Hurst, the fractal concept was unknown. But he set about to mathematically describe this feature by creating the method of rescaled range analysis. The range of time-series data R is defined as follows:

$$[Y_{max} - y_{min}] = R(\Delta t)$$

where Y is the time data and max indicates maximum value during time interval Δt Range R and Min is the minimum value during the same time interval. The difference between the maximum and minimum values during time interval Δt is the range $[\Delta t]$. Then, find the standard deviation $S(\Delta t)$ over the same time interval Δt. Hurst defined the rescaled range as R/S and defines the Hurst exponent as

$$\frac{R(\Delta t)}{S(\Delta t)} = c\Delta t^{H} \tag{8.2}$$

Equation 8.2 more explicitly analyzes the long-term memory of fractal time. Furthermore, the power law characteristic of fractal appears once again in this equation where the value of H may be related to the fractal D.

In summary, the power law characteristic of fractal time was analyzed by various methods in this chapter. First, to treat it as a spatial object; second, to analyze its frequency spectrum; and lastly, by means of the Hurst dimension. Fractals will be applied in the next chapter to help quantify the blood flow variation in the left ventricle.

Problems

1. While sitting, measure your heart rate from your pulse during a 1, 2, 5, 10, and 20 min periods. Graph the log heart rate versus log time. Did you find a linear log–log graph? If so, measure the slope.
2. Equation 8.1 describes the frequency content typical of fractal time. Using your data in Problem 1, show that the frequency content is consistent with Eq. 8.1 for heart rate variation.
3. Generate a random time series using the Gaussian random function of your software such as randn for MATLAB to reproduce Fig. 8.4. Let each random number correspond to a time i. Plot randn [i] versus i.

4. Generate a random walk (Brownian) motion time series by iterating the random number generator in Problem 3. Use numerical integration. For N points RW(i) = RW(i−1) + randn$_i$ That is, add a Gaussian random number to the current point to get the next point. Then, graph RW(i) versus i. You should notice the longer periods present in the random walk, as suggested by Fig. 8.6 or Fig. 8.2.

5. Optional problem. Listen to some of your favorite music and count the number of "beats" over 1, 2, 3, 4 minutes. Do the same for what is generally considered bad music. Summarize both in a beats versus time interval plot. You should find that good music has a log–log relationship as we know that fractal music sounds better to the human auditory system for which the cochlea is a fractal spiral geometry.

References

Frank, S. A. (2009). The common patterns of nature. *Journal of Evolutionary Biology, 22*(8), 1563–1585. https://doi.org/10.1111/j.1420-9101.2009.01775.x

Goldberger, A. L., Amaral, L. A., Hausdorff, J. M., Ivanov, P., Peng, C. K., & Stanley, H. E. (2002). Fractal dynamics in physiology: Alterations with disease and aging. *Proceedings of the National Academy of Sciences of the United States of America, 99*(Suppl 1), 2466–2472. https://doi.org/10.1073/pnas.012579499

Hurst, H. E. (1951). Long-term storage capacity of reservoirs. *Transactions of the American Society of Civil Engineers, 116*(1), 770–799. https://doi.org/10.1061/TACEAT.0006518

Katz, M. J. (1988). Fractals and the analysis of waveforms. *Computers in Biology and Medicine, 18*(3), 145–156. https://doi.org/10.1016/0010-4825(88)90041-8

Mandelbrot, B. B. (1983). *The fractal geometry of nature* (Updated and Augm. ed.). W.H. Freeman.

Mandelbrot, B. B., & Mandelbrot, B. B. (1982). *The fractal geometry of nature* (Vol. 1). WH freeman.

Peitgen, H.-O., Saupe, D., & Barnsley, M. F. (1988). *The science of fractal images*. Springer.

Peitgen, H. O., Fisher, Y., Saupe, D., McGuire, M., Voss, R. F., Barnsley, M. F., … Mandelbrot, B. B. (2012). *The science of fractal images*. Springer.

Rapp, P. E., Bashore, T. R., Martinerie, J. M., Albano, A. M., Zimmerman, I. D., & Mees, A. I. (1989). Dynamics of brain electrical activity. *Brain Topography, 2*(1), 99–118. https://doi.org/10.1007/BF01128848

Biomedical Signal Analysis and Fractal Time

9

9.1   Fractals and Time Signals

The previous chapter on Fractal time series may be familiar to those with some signal processing background. Some comparisons between fractal time and information signals are listed here (Katz, 1988).

A. Time waveforms generally increase in one dimension monotonically with time. It is not usual for information signals to cross back on themselves. So, time waveforms do not define a planar object as a fractal might.
B. Signal waveforms are simple in that only one parameter is required to describe them. For example, a line, an exponential pulse, sinewave, and square pulse require only a single or few parameters to describe them.
C. Nonideal waveforms may be nonparametric or require a complex set of defining parameters.
D. Nonparametric waveforms must often be described by means of feature extract methods like Fourier, peak, or wavelet models.
E. Other models are not needed to describe fractal time.

Since a chaotic process is deterministic, its time series may be fractal and the existence of fractal properties like self-similarity is evidence of chaos. Information signals are deterministic as well, so it can be expected that some time signals are fractal and possess a fractal dimension D. Signal processing theory may apply this knowledge by attempting to find a dimension in signal with noise, as is the usual activity of signal processing. Therefore, the existence of a D value indicates with some level of certainty that a time signal possesses fractal properties and is likely to be deterministic Hence, the D value suggests with some probability the existence of a deterministic process.

The information gained by the D value is that of the scaling property of the waveform but does not go beyond this. Qualitatively, the value of D indicates the complexity of a signal and the system that produced it. Hence, care needs to be exercised

G. Drzewiecki, *Fundamentals of Chaos and Fractals for Cardiology*,
https://doi.org/10.1007/978-3-030-88968-5_9

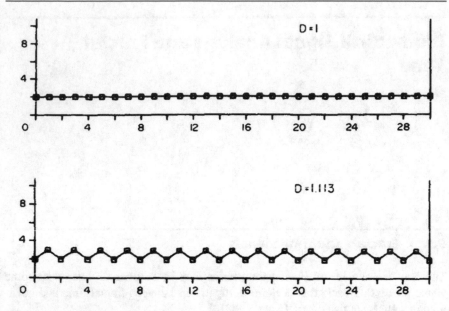

Fig. 9.1 The line (top) and triangle (bottom) waveforms are shown as value versus time point number. (Reproduced from Katz, 1988)

while applying Fractals to time signals in concluding beyond what information is really gained from D. As for a geometric object, it was shown that fractal geometry fills a space. This property is defined as the "space-filling" property of the fractal geometry. The space-filling concept may be applied to time waveforms as well. Examine the fractal dimension of a few well-known time signals. For example, a DC and triangle waveforms are shown in Fig. 9.1.

The fractal dimension values for some typical time waveforms are:

1. The straight line: D = 1.0.
2. The triangle wave: D = 1.113.
3. The square wave: D = 2.0.

9.2   Finding Fractal D of Time Signals

In summary, the value of D increases with the space-filling property of a signal waveform, and since D reflects the possible states of a waveform, it also indicates the relative complexity of the system that produced the waveform. Following this concept, Katz provides a convenient method of finding D from time signal waveforms. The method is now reviewed. The Fractal is redefined for a planar object as:

$$\text{Planar fractal Law is } D = \frac{\log L(r)}{\log d(r)} \tag{9.1}$$

where L is the data length, r = increment in length, and d = planar diameter of the data. L can be found from the sum of distance between all data points where the distance is the vector distance between adjacent data points distance such as two points $(y_1.t_1)$ and (y_2,t_2).

Equation 9.1 is generally applicable to any time series of data. It may therefore be employed as a simple test for deterministic or random data. Katz also provides a method for testing for a deterministic signal, as addressed in the next section.

9.3   Procedure to Test for Random Signals Using the Fractal D

Given a signal $S(t)_N$ that contains N- points, find its Fractal D as $D\{S\}$.

1. Generate a random walk time series of N-points R_N.
2. Calculate D for the random walk = D[R].
3. Perform several trials for D[R].
4. Create a histogram of all trials of D[R] for each random walk generated.
5. Apply the statistical test of ANOVA to find if the $D\{S\}$ lies within the D values of the random walk to obtain the p-value for D[S].
6. If P > 0.01, then the data S are not random within 1% certainty.

It is important to keep in mind the fact that true random time series are very difficult to attain in reality. Additionally, the usual software algorithms that generate random number series for computational studies are also not random. Computer algorithms produce the standard computational methods that provide random numbers. While these algorithms may provide a good approximation to random, by virtue of the use of an algorithm, these random generators are actually deterministic such that knowledge of the algorithm permits the prediction of the numbers that have been created. Random number generator software is more likely to generate Chaos than true random numbers.

Hence, it is more rigorous to apply other tests of Chaos discussed earlier in this book in addition to the above Fractal D test for random. Fractal D testing might be viewed as a quick screen before more rigorous testing is performed.

The last subject of testing for random is that of required data length. Simply put, how much data is enough to accurately determine the fractal dimension of a time series. Unfortunately, there is no good formal criterion. Instead, an experimental method may be applied as introduced by Kaplan and Goldberger (1991). In finding the fractal D, any of the methods described so far may be applied to a time series with equal validity. First, choose an arbitrary data length and calculate the value of D. Next, repeat the D value calculation for an increased amount of data period. This process of increasing the data duration and D value calculation is continued until the maximum amount of data has been used to find D. After completing the D values for the entire data length, D is then available as a function of time duration T. The D {T} is then examined graphically. Usually, it will be observed that a steady value of D is attained over a range of T. Graphically this will appear as a plateau of the data.

This approach is analogous to the steady-state solution of a differential equation. The value of D that is independent of T is then chosen to be the correct D. Further confidence in the observation of D can be obtained by applying the use of statistics to the D {T} curve. You would then choose to use the D value that varies less than a standard deviation D over the other range of T. The sensitivity of D to other data sampling parameters may also be explored in a similar way. For example, the data sample rate Δt may be observed graphically as the function D {Δt}. The same procedure is applied to find the correct Δt. D (Δt) is observed for values of D that are insensitive to Δt. Then, Δt is chosen as the value that results in an invariant D.

9.4   Fractals in Cardiology or "Nonlinear Cardiology"

The topic of nonlinear dynamics applied to cardiology was presented earlier in the "nonlinear cardiology" chapter. In this chapter, we return to nonlinear cardiology by applying the use of fractals. Fractal methods to analyze the electrocardiogram (ECG) were introduced by Goldberger and West (1987). Goldberger's studies were primarily in the frequency domain so that the methods described in Chap. 8 were employed. Beginning with the QRS wave of the ECG, the QRS wave was analyzed in the frequency domain by means of the fast Fourier transform (FFT). FFT for a normal human subject is shown in Fig. 9.2. Notice that the log–log plot shows an inverse frequency relationship with magnitude so that $ECG \propto 1/f^{\beta}$.

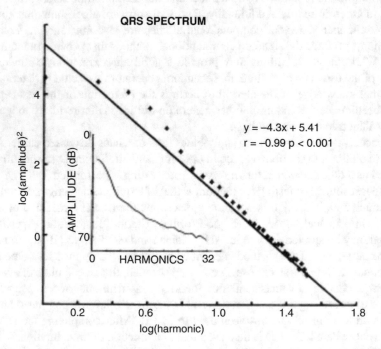

Fig. 9.2 Log–magnitude [ECG]–LogF plot of the Fourier ECG. (Reproduced from Goldberger & West, 1987)

Goldberger found that normal QRS wave yields an inverse relationship between log magnitude [QRS] and log frequency. In accordance with the inverse frequency formula of Chap. 8, it can be concluded that the QRS wave of the ECG is fractal time. Goldberger further suggested that since the QRS wave is created from a branching Purkinje network it might be expected that a fractal branching network might generate a time signal that is also fractal. Therefore, the concept of the fractal may represent the health of a fractal network. This approach is referred to as nonlinear or fractal cardiology. The fractal concept as applied to the ECG analysis provides a new way of gaining additional diagnostic information about the cardiovascular system and heart. Goldberger provides a few new diagnostic insights. To illustrate some fractal cardiology, a few abnormal ECG waves are shown in Fig. 9.3.

Figure 9.3 shows that high values of D indicate an abnormal ECG waveform. Goldberger suggests that a loss of high-frequency magnitude of the QRS wave may be abnormal.

A cardiac infarct is likely indicated by a loss of fractal characteristics. He also finds that a narrowing of the QRS spectrum into a single frequency indicates tachycardia that often progresses to ventricular fibrillation. In summary, the value of the fractal concept applied to the ECG yields more predictive indications as opposed to just an indication of current cardiac conditions.

The ECG may be examined for longer time periods to include several heart beats where it is evident that a much lower frequency magnitude–frequency spectrum exists. A sample is shown in Fig. 9.4.

This frequency range is also referred to as the heart rate variability spectrum. Observe that this spectrum has an inverse frequency relationship. Hence, heart rate variation is a fractal time series. There are also some large magnitude peaks in the spectrum. The middle peak is due to sympathetic and parasympathetic regulation. The right-side peak is due to the respiratory rate. Observations of the heart rate variability spectrum fractals reveal the feature of period doubling or bifurcations. For example, Goldberger and West (1987) report a 2X and 4X rhythms that correspond with the condition of AV block. Due to the fact that the cardiac rhythm regulation system is nonlinear but with a long time delay, it is expected that heart rate variability should be a chaotic time series. Kaplan and Goldberger (1991) propose that a chaotic cardiac rate is normal. Cardiologists refer to the normal variation in heart rate as sinus arrhythmia. They further suggest that loss of chaos indicates pathology. Loss of heart rate variation was found to occur in heart failure patients (Goldberger et al., 2002). Hence, the value of observing the fractal variability of the heart rate may lie in its ability to predict patient outcome and evaluate the success of treatment in a more timely manner. Recalling from the earlier Chap. 2, it was shown that two nonlinear oscillators interact to result in a chaotic output. While cardiologists prefer to define 2 to 1 rhythms as heart block, similar kind of rhythm can be produced by the interaction of coupled nonlinear oscillators. It is proposed by Goldberger and West (1987) that heart block rhythm may simply be a result of the interaction of the SA and AV nodes of the heart. Hence, the nonlinear cardiologist possesses an alternate view of arrhythmia. Nonlinear cardiology may provide a new perspective on a major research area of the cardiovascular system. For example, the nonlinear

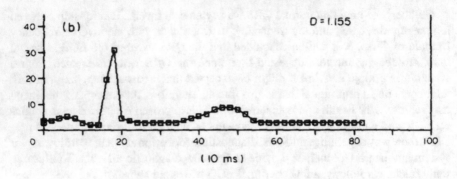

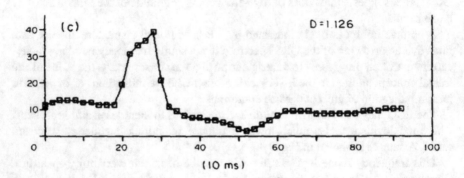

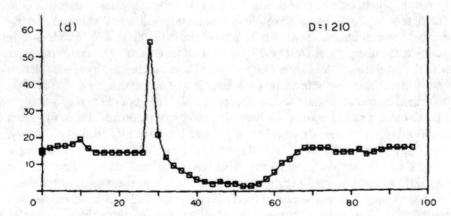

Fig. 9.3 Three ECG waves with their Fractal dimension D. (**b**) is normal; (**c**) is a patient with bundle branch block; (**d**) is the ECG of a myocardial infarct. (Reproduced from Katz, 1988)

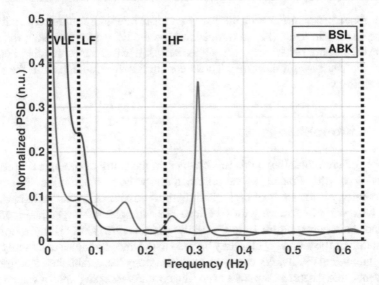

Fig. 9.4 Sample frequency spectrum of a normal beat-to-beat heart rate variation. BSl denotes the basal heart rate data. ABK denotes an autonomic blockade. (Reproduced from Rosenberg et al., 2020)

approach views cardiac arrest as a state of the nonlinear oscillator that is simply no output. Hence, a nonlinear approach to treatment might be to prevent conditions that lead the oscillator to the zero output state that is cardiac arrest.

9.5 Measuring Heart Rate Variability

Realizing the value of observing heart rate variability, it is important to discuss its proper measurement. At first approach, it may be that an ECG waveform may be digitized and then analyzed by performing the fast Fourier transform (FFT) on the time-series data. Unfortunately, the FFT may introduce artifacts into the spectrum. An artifact is known as aliasing (Semmlow, 2017). A second occurs due to the FFT algorithm that expects data to be uniformly sampled. Due to the beat-to-beat variation in heart period, a uniform sample does not occur. A more accurate frequency spectrum procedure should be as follows:

1. Detect the R- waves of the ECG by using a peak detection method.
2. Convert the beat-to-beat time into time interval between R's; this results in the new time series of beat-to-beat period versus time.
3. Convert period into heart rate by calculating the inverse periods versus time.
4. Low pass filter the heart rate time series at 1 Hz to remove any erroneous low-frequency artifact.
5. Lastly, resample the filtered heart rate time series at a uniform rate to obtain the final heart rate time series. The FFT is then applied to the resampled heart rate data.

The above procedure forces the focus on heart rate data to the frequencies less than the heart rate itself. Those frequencies are the heart rate variation, the neural regulation, and the respiratory rate. Other very low frequencies are also known to exist in the heart rate time series. Those are studied by researchers in the area of chronobiology.

9.6   Chronobiology

The topic of chronobiology ends this chapter on the study of biological time series by means of the Fractal as an alternative method to examine time series. Chronobiology is the study of biological rhythms outside of fractal variations, especially those variations of very long periods For example, in human the blood pressure and body temperature follow a daily variation in value. The physiological information of the human cycles may be used to enhance therapies and more closely time pathological events. For example, chronobiology has established that most cardiac events, like a heart attack, will occur during early morning upon awakening and at the time of dinner (Vitaterna et al., 2001). Chronobiology helps to enhance chemotherapy if it is administered following the patient's chronobiology (Bjarnason, 1995). Most human physiology variables have been observed over various fixed time periods. So chronobiology differs from the fractal method in that the FFT is not employed. Instead, the chronobiology model is a cosine function as follows for any physiological variable $x(t)$:

$$x(t) = A\cos(\omega t + \varphi) + B \tag{9.2}$$

where A is the amplitude of variation, ω is the angular frequency, t is the time, φ is the phase, and B is the mean level.

Nonlinear data estimation methods are then used to vary the above parameters so that Eq. 9.2 is a best fit with the data time series. After performing this fitting procedure, some basic descriptive measures result are defined below:

Mesor = A/2.
Maximum = Acrophase
Minimum or = acrometron

Some common chronobiological rhythms are as follows:

1. Circadian = 24 h
2. Infradian >24 h
3. Ultradian <20 h
4. Circaseptian = 7 days
5. Circa annual = 1 Year

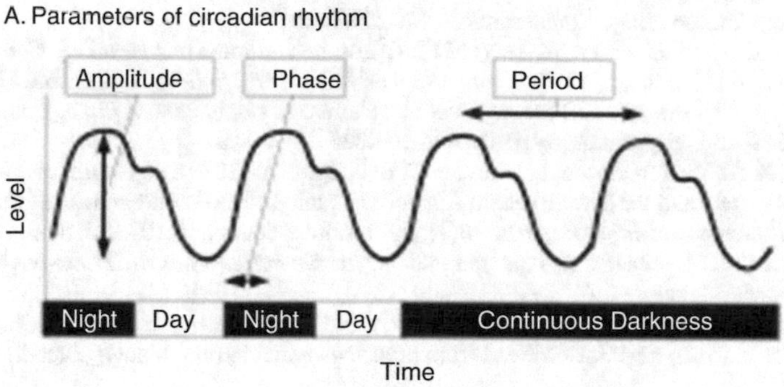

A. Parameters of circadian rhythm

Fig. 9.5 Example of human physiological variables as a circadian cycle. The light region is day-time, and the dark region is night. (Reproduced from Vitaterna et al., 2001)

Figure 9.5 provides an example of a typical human physiological variable as a circadian rhythm. The time of acrophase is the maximum amplitude. Most of the cardiopulmonary variables maximize in the mid-afternoon.

Coinciding with the cardiopulmonary measures, it is found that physical and cognitive tests are best in the mid-afternoon.

Problems

1. Explain why the QRS waveform of the ECG exhibits a fractal frequency relationship.
2. When studying the frequency spectrum of heart rate variation, what are the physiological systems responsible for low-frequency variation?
3. Is the heart rate variability spectrum closer to the random walk or that of random noise?
4. Compared to fractals, what is the value of chronobiology measurements after taking frequency content into account?
5. What does the fractal dimension indicate about the ability of a physiological system to adapt?

References

Bjarnason, G. A. (1995). Chronobiology implications for cancer chemotherapy. *Acta Oncologica,* *34*(5), 615–624. https://doi.org/10.3109/02841869509094037

Goldberger, A. L., Amaral, L. A., Hausdorff, J. M., Ivanov, P., Peng, C. K., & Stanley, H. E. (2002). Fractal dynamics in physiology: Alterations with disease and aging. *Proceedings of the National Academy of Sciences of the United States of America, 99*(Suppl 1), 2466–2472. https://doi. org/10.1073/pnas.012579499

Goldberger, A. L., & West, B. J. (1987). Fractals in physiology and medicine. *The Yale Journal of Biology and Medicine, 60*(5), 421–435. Retrieved from https://pubmed.ncbi.nlm.nih. gov/3424875 https://www.ncbi.nlm.nih.gov/pmc/articles/PMC2590346/

Kaplan, D. T., & Goldberger, A. L. (1991). Chaos in cardiology. *Journal of Cardiovascular Electrophysiology, 2*(4), 342–354. https://doi.org/10.1111/j.1540-8167.1991.tb01331.x

Katz, M. J. (1988). Fractals and the analysis of waveforms. *Computers in Biology and Medicine, 18*(3), 145–156. https://doi.org/10.1016/0010-4825(88)90041-8

Rosenberg, A. A., Weiser-Bitoun, I., Billman, G. E., & Yaniv, Y. (2020). Signatures of the autonomic nervous system and the heart's pacemaker cells in canine electrocardiograms and their applications to humans. *Scientific Reports, 10*(1), 9971. https://doi.org/10.1038/s41598-020-66709-z

Semmlow, J. (2017). *Circuits, signals and systems for bioengineers: A MATLAB-based introduction*. Elsevier Science.

Vitaterna, M. H., Takahashi, J. S., & Turek, F. W. (2001). Overview of circadian rhythms. *Alcohol Res Health, 25*(2), 85–93. Retrieved from https://www.ncbi.nlm.nih.gov/pubmed/11584554

Deterministic Fractals

10

10.1 Introduction

Up to this point, fractals have been applied to study natural geometries and biological time waveforms. Also, a chaotic time series is generally fractal. In Chap. 2, the limited population equation was supplied in its differential equation form and then converted into a numerical solution form constructed by using Euler's method (Ascher & Petzold, 1998). This equation was further arranged to its iterate form of solution (the logistic equation). The logistic equation was shown to generate a chaotic time series. In the most recent fractal chapters, it was shown that time series such as the logistic equation are also fractal time. In this chapter, you will see that there are other nonlinear iterates that generate chaotic time series and are fractals. These forms of iterate equations are more purely mathematical in form and may not be related to natural fractal geometries or dynamic systems, but they possess the property of being fractal. These mathematical fractals are defined as the deterministic fractals.

This chapter introduces the deterministic fractal. The deterministic fractal is not only a mathematical field of study. This chapter will also show some applications of deterministic fractals to model geometry such as that found in images where there is a large amount of self-similarity or objects that repeat themselves.

10.2 The Julia Set

Some deterministic fractals have been defined by mathematicians as sets of objects. The Julia set of fractals is a set of deterministic fractals (Peitgen & Richter, 1986).

The Julia set is created from the following iterate function F.

The variable z is a complex number $z = x + iy$, where "i" denotes the imaginary number.

© The Author(s), under exclusive license to Springer Nature Switzerland AG 2021
G. Drzewiecki, *Fundamentals of Chaos and Fractals for Cardiology*,
https://doi.org/10.1007/978-3-030-88968-5_10

$$F(z) = z^2 \tag{10.1}$$

Expanding Eq. 10.1 reveals Eq. 10.2:

$$z^2 = x^2 - y^2 + i2xy. \tag{10.2}$$

Equation 10.2 is then iterated to find z.

Starting with the initial value of z, iterate Eq. 10.2 for N- iterations and search for any of these three conditions of Eq. 10.3:

1. For $0 < |z| < 1$ for $\lim^0$
2. $|z| > 1$ for $\lim \infty$ (10.3)
3. $|z| = 1 = $ Chaos

The Julia set is defined as condition number 3 of chaos and is a fractal.

The values of z that result in chaos can be shown graphically as an X–Y plot after converting z to its real and imaginary form. A more general form of the Julia set fractal can be created by adding an imaginary constant, c, to Eq. 10.1 as follows:

$$F(z) = z^2 + c \tag{10.4}$$

Equation 10.4 is then iterated as before to find the three conditions in Eq. 10.2. Figure 10.1 illustrates four sample Julia sets for Eq. 10.4 using four different values of C = [−0.1, 0.25, 0.12, −0.13].

The axis is +/−2 for both x and y. White data points indicate convergence to zero. The black data points indicate solutions that tend to infinity.

Fig. 10.1 The Julia sets for Eq. 10.4 with C = [−0.1, 0.25, 0.12, −0.13] corresponding with fractals [a, b, c, d]. (Reproduced from Peitgen et al., 1988)

Recall back to the nonlinear dynamics portion of this book. The Julia set examples provided in Fig. 10.1 are the attractors of Eq. 10.4 for different values of c. The black and white images are created by applying the definition that black points are a result of an unstable solution that is infinite, which is condition 2 of Eq. 10.3. The white areas are solutions that trend to zero. The borderline between black and white is the remaining solution that is chaos. The chaotic solution is therefore the boundary between a zero and infinite result. Thus, besides providing a pure mathematical fractal, the Julia sets also clearly demonstrate that chaotic strange attractor can be a fractal.

While the math to generate the deterministic fractal is quite simple, Eq. 10.3 must be iterated a large number of times to produce the images as of Fig. 10.1.

It is therefore best to approach this calculation in a computationally efficient manner. To begin the fractal computation, the maximum number of graphic points needs to be defined. For example, the graphics plot is limited to N x N total points or pixels. So, at the worst case, we must iterate Eq. 10.4 at every pixel to observe its stability. These points may be cut in half by using the inverse iteration method (Peitgen et al., 1988). This method follows from the fact that the calculation consists of imaginary numbers that have magnitude and phase. Starting from any initial value z_0 , the square root may be used to find the points from which z_0 was determined. They are the two points $\sqrt{z_0}$ $\theta, (\theta / 2 + \pi)$. Thereby, one known fractal point immediately provides two others, cutting the total calculations in half.

Another way of reducing the calculations is to minimize the number of iterations for each pixel. Notice that the Julia set fractals of Fig. 10.1 indicate an unstable value as a black pixel. Searching for $z \rightarrow z \lim_{z \to \infty} \Rightarrow \infty$ would have no limit of iterations. Instead, realizing that the instability occurs quickly after a few iterations, it can be chosen that when $|F(z)| > 3$ for N = 30 iterations or less, that z is unstable. This quickly limits the total number of iterations needed for a deterministic fractal. You may have seen the more beautiful Julia set images produced elsewhere (Peitgen & Richter, 1986) that also color each pixel according to the number of iterations needed to find when the magnitude of z is greater than 3.

The Julia set is part of a larger set of fractals. If we continue to examine Eq. 10.4 for different values of c and initial condition z = 0, it will be discovered that some values of c are nondivergent. These stable Julia set fractals are themselves a set that is defined as the Mandelbrot set after the originator of the fractal himself (Mandelbrot, 1982). The Mandelbrot set is provided graphically in Fig. 10.2.

The more efficient computational methods described above were used to locate those Julia sets that are connected. All Julia sets were located from iteration of Eq. 10.3. The most cumbersome aspect of creating a full fractal image is the necessity of computationally exploring the many conditions for the purpose of locating the values that result in chaos by iterating each separately.

Graphical iteration is another means by which fractal objects can be created. For example, the Sierpinski fractal is generated by implementing a simple triangle object repeatedly until the full triangle is created. The iteration rule and Sierpinski fractal are provided in Fig. 10.3 (Peitgen & National Council of Teachers of Mathematics, 1999).

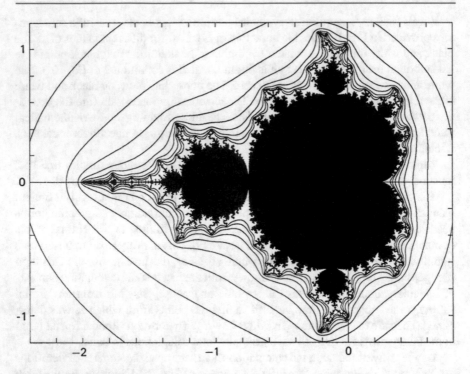

Fig. 10.2 The Mandelbrot set is shown. The real and imaginary values of c that provide stable Julia sets are shown as black points

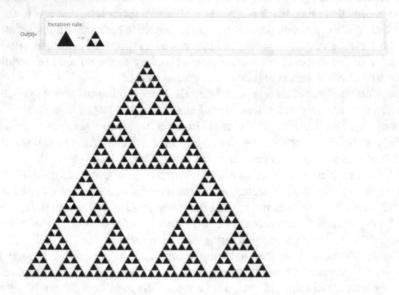

Fig. 10.3 The Sierpinski fractal generated using the above iteration rule

Notice the self-similarity of the object. The dimension value is D = 1.584. It is important to understand that the fractal is generally created by means of some iterative procedure. Two types of iterations have been shown in this chapter, the mathematical function and the geometric or graphical scaling rule iteration of the Sierpinski triangle. Very complex objects were created by these iterative methods. Yet, in both methods shown here, the iterative rule was very simple. In the context of bioengineering, the iteration of a simple rule is important to biological structures. In Chap. 7, the structures of the human lungs and vascular system were shown as two biological examples of fractals. These structures were created by means of the method of dichotomous branching, the branch of a single tube into two smaller tubes. As a result, the lungs and vasculature are fractal (Nelson et al., 1990). These geometrical fractal iterates are defined under the more general geometric algorithm defined by Prusinkiewicz et al. (2013), also known as the L-system.

To further illustrate the concept of biological construction via the fractal iterate, observe some examples of plant geometry as generated by means of the L-system (Prusinkiewicz et al., 2013).

A plant example is provided in Fig. 10.4 that was generated by an L-system.

The L-system model is well suited to modeling plant structures as it incorporates plant growth structures into its generation. These again are very simple structures created by a minimum of graphical movements such as angle turns or forward motions.

Some mathematical iterates have also been found to generate objects surprisingly similar to cellular objects as shown in Fig. 10.5.

Each object is shown with its corresponding fractal function.

Fig. 10.4 Fractal plant model generated using the L-system method. (Reproduced from Prusinkiewicz et al., 2013)

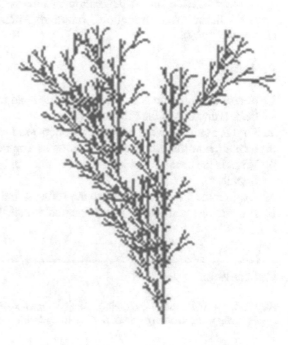

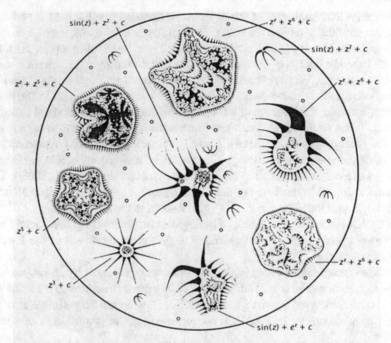

A microscopic view of some biomorphs and their respective generating functions

Fig. 10.5 Cellular objects created from mathematical iterations

Another flexible fractal generation method is the iterate function system that applies linear transformation functions iteratively to re-create images (Dewdney, 1989).

Problems

1. Iterate the equation $X_{n-1} + X_{n-2} = X_n$ by hand to produce a series of nine numbers. (Hint: this is the Fibonacci series.)
2. Find the ratio between adjacent numbers produced in Problem 1.
3. Explain how the Fibonacci series models a spiral geometry.
4. What is the process that is used in common to both deterministic and geometric fractals?
5. Explain the difference between the Julia set and Mandelbrot set.
6. Provide one example of a mathematical iterate that relates well to biology.

References

Ascher, U. M., & Petzold, L. R. (1998). *Computer methods for ordinary differential equations and differential-algebraic equations.* Society for Industrial and Applied Mathematics.

Dewdney, A. K. (1989). Computer recreations. *Scientific American, 261*(1), 110–113. Retrieved from http://www.jstor.org/stable/24987330

Mandelbrot, B. B. (1982). *The fractal geometry of nature.* W.H. Freeman.

Nelson, T. R., West, B. J., & Goldberger, A. L. (1990). The fractal lung: Universal and species-related scaling patterns. *Experientia, 46*(3), 251–254. https://doi.org/10.1007/BF01951755

Peitgen, H.-O., & National Council of Teachers of Mathematics. (1999). *Fractals for the classroom: Strategic activities* (Vol. 3). Springer.

Peitgen, H.-O., & Richter, P. H. (1986). *The beauty of fractals: Images of complex dynamical systems.* Springer.

Peitgen, H.-O., Saupe, D., & Barnsley, M. F. (1988). *The science of fractal images.* Springer.

Prusinkiewicz, P., Lindenmayer, A., Fracchia, F. D., Hanan, J., & Krithivasan, K. (2013). *Lindenmayer systems, fractals, and plants.* Springer.

Fractals and the Cardiovascular System

11

11.1 Myocardial Blood Flow

Beginning with this chapter, some applications of nonlinear dynamics and fractals to physiological research are introduced and cardiovascular system research will be presented.

Fractal time was shown earlier to apply well to the heart rate time series. Additionally, the QRS wave of the ECG was also shown to be a fractal time waveform. Since the ECG arises fundamentally from a fractal network of excitable cells, it was suggested that a fractal signal network should yield a fractal time waveform (Goldberger et al., 1985). In this chapter, the focus is on the fractal geometry of the vascular flow network, particularly the study of blood flow in the heart. Researchers of myocardial ischemia have found that tissue damage due to loss of blood flow tends to be localized in the heart to specific regions of typically low flow (Hoffman, 1995).

Bassingthwaighte et al. (1989) used the fractal concept to analyze the variation of blood flow throughout the left ventricular muscle. The experimental approach of the Bassingthwaighte study was to inject radioactive particles into the main coronary artery of a canine heart. These particles would then distribute throughout the myocardial tissue according to flow patterns and the coronary network. The animal was then sacrificed and the heart removed and sectioned into N-pieces. Blood flow to each piece of the myocardium was determined by measuring the radioactivity of the piece. The flow to each piece was then analyzed in terms of its relative dispersion, where RD(F) = relative dispersion of flow, F. RD was calculated from the ratio of standard deviation of RD(F) mean (F) for all pieces. All RD values were plotted in a histogram of flows as shown in Fig. 11.1.

The histogram revealed a surprising result in that the RD histograms changed depending on the number of tissue sample pieces that are analyzed. Assuming that the coronary artery system is a branching network, the dispersion data were reanalyzed using fractal analysis. Assuming a fractal, it would be expected that the RD measurement should depend on the sample size N. The fractal geometry law was

© The Author(s), under exclusive license to Springer Nature Switzerland AG 2021
G. Drzewiecki, *Fundamentals of Chaos and Fractals for Cardiology*,
https://doi.org/10.1007/978-3-030-88968-5_11

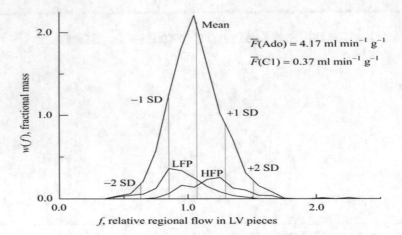

Fig. 11.1 Histogram of relative dispersion of blood flow in the heart (F). Total number of pieces, N, examined is indicated on each curve. (Reproduced from Bassingthwaighte et al., 1989)

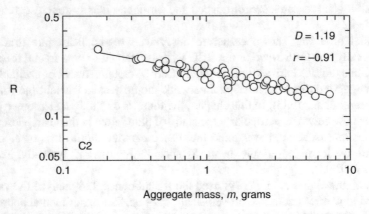

Fig. 11.2 Fractal law log–log plot of RD and sample size. (Reproduced from Bassingthwaighte et al., 1989)

applied to replot the data in terms of log RD versus Log N. The results are shown in Fig. 11.2.

Fractal analysis provides a clear picture that the log–log relationship is linear and therefore supports the fractal hypothesis. The data follows a log–log linear fit with a correlation of 0.992. The resulting fractal relationship was found to be RD = $0.129N^{0.79}$. Therefore , the fractal dimension of the coronary circulation is D = 1.2. Now, knowing the fractal model of blood flow, we can return to solve the problem of regions of low blood flow. For example, find the flow near a single myo-cardial capillary. Using a capillary mass of (0.2–1 mg) in the fractal equation yields an RD = 60%. This means that capillary flow may vary by +/−60% of its mean value. In summary, it is found that any total flow reduction that exceeds this 60% value can result in a flow reduction to zero in the region of a capillary. This study reveals that the coronary circulation design is borderline in its design to consistently provide blood flow to every capillary. In addition, this study has shown that the variation in blood flow to the ventricular tissue follows a fractal scaling law. This observation is consistent with the fact that the microcirculation of the heart is a fractal branching network. The study of branching network structures provides a more detailed picture of microvascular design.

11.2 Microcirculation Branching Structures

Much of the knowledge of vascular microcirculation structure comes from microscopic studies of the bat wing circulation (Wiedeman et al., 1981). Figure 11.3 illustrates one of the common branching schemes called dichotomous branching. In this network, each vessel branches into two others. This process continues until a capillary is reached.

Figure 11.3 also shows the centripetal vessel numbering scheme. In this number system, each branch results in a higher number. The choice of numbering system is mostly arbitrary, but the centripetal numbering in Fig. 11.3 provides some idea of vessel diameter and length by the branch number. The dichotomous branching system is most commonly found in the circulation of the human gut. Alternatively, the Strahler branching system of Fig. 11.4 is found in the coronary circulation.

In the Strahler system, vessels that converge increment to a higher number.

Now that some vascular network structures have been defined, it is important to analyze their ability to deliver blood flow to the capillary bed. The branching structure needs to be quantified. First, define the branch ratio RB_α where α I is the branch number. Let the number of branches at the current branch be N_i, and the number of branches at the next branch level is N_{i+1}. RB_i is defined as $RB_i = \dfrac{N_i}{N_{i+1}}$. RB may be used to measure various vascular beds. In general, it has been found that the branch number α is proportional to the log of the number of vessels on average. This suggests the power–law relationship of $N_i = N_1 \ xRB^{i-1}$.

The branching ratio measure may also be analyzed by applying statistical studies of RB. For example, branching histograms can reveal the probability of branching as seen in Fig. 11.5.

Fig. 11.3 Bat wing microcirculatory dichotomous branching network. Numbers indicate the branching number according to the centripetal system. (Reproduced from Wiedeman et al., 1981)

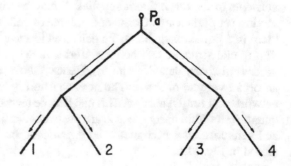

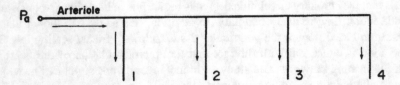

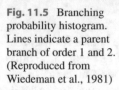

Fig. 11.4 Example of the Strahler branching system. (Reproduced from Wiedeman et al., 1981)

Fig. 11.5 Branching probability histogram. Lines indicate a parent branch of order 1 and 2. (Reproduced from Wiedeman et al., 1981)

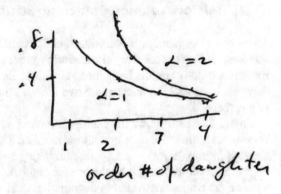

The data of Fig. 11.5 reveal that similar numbered vessels are more likely to branch than those that have very different numbers. While these relationships can describe the structure of various branching networks, they help to analyze the hemodynamic function to discover the physical reason of how networks form the structures that are observed. In addition to RB, there are other measures that have been defined. For example, there are the length ratio, LR, and the diameter ratio, DR. For fractal networks, these measures also follow the power law as

$$L_i = L_1 * (LR)^{i-1} \text{ and } D_i = D_{1*} (DR)^{i-1}$$

11.3 Vascular Design Principles

While the fractal structure is a well-defined structure that permits the distribution of nutrients to the organ that it supplies, it may be viewed as a system that performs mixing very efficiently. Consider a volume of cells that require oxygen. If we place them in a liquid, oxygen may be delivered to the cells using some form of mixing. The fractal structure can be viewed as a network that performs the same mixing process in a very organized means. Hence, the fractal network mixes nutrients into an organ volume of tissue. The next question is how much energy does a fractal network cost for its function. This can best be answered by examining the two basic quantities of fluid energy, which are pressure and flow velocity. The fluid pressures and velocities have been measured throughout the human vasculature and are presented in Fig. 11.6.

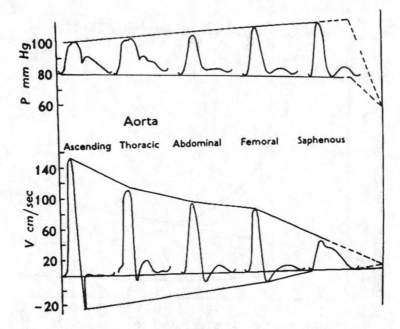

Fig. 11.6 Vessel pressure and velocity throughout the vascular system as indicated by anatomical position. The microcirculation is on the right side of each graph. Aorta is on the left. (Reproduced from Noordergraaf, 1978)

Moving from left to right in Fig. 11.6 begins at the heart and continues through large vessels then to the microcirculation. In both the graph of pressure and velocity, it is seen that values trend toward zero once the microvessels are reached. Since the product of pressure and velocity represents fluid energy, we find that almost all of the energy of the heart is lost by the time it reaches the vascular beds. Hence, from an energetic point of view the vascular system is an energy loss system. Physically the energy is lost from heat due to flow. It is initially disappointing that the fractal structure might be a perfect solution for transport physiology.

It is clear from the above that the fractal serves as a valid model of vascular structure. By combining the structural model with vascular function, it is possible to design a vascular network from start. To do this design, it is necessary to first define the important functional parameters for the vascular network and its physiology. Basic hemodynamics at the capillary level requires that blood pressure and flow velocity fall within a specified range for proper nutrient transport. An alternate approach is to calculate the energy consumption of the vascular network at the input vessel to the entire network. Then, find the total flow input resistance as R_{input}.

Total input power is obtained as Power = R_{in} x Q^2, where Q is flow input. Total vascular input power can then be computed in terms of the fractal branching ratio RB and N the number of branches. Wiedeman et al. (1981) performed the power calculations for the basic dichotomous network. Some fundamental findings emerge as shown in Fig. 11.7.

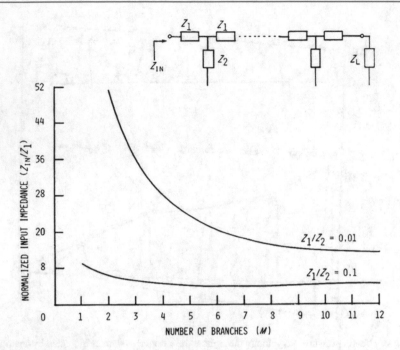

Fig. 11.7 Input impedance to a vascular network computed as a function of total branch number N. (Reproduced from Wiedeman et al., 1981)

Since power is proportionate to the input impedance Zin, it is observed that power and therefore energy requirements of a vascular network decrease with more branching. It can be concluded that more branching is a more energy-efficient system of transport in terms of energy utilized to move material. In general, nature tends toward using branching in most plants and animals for transport as the fractal is also likely the most efficient design. Following a similar analysis, Wiedeman et al. (1981) next examined the most efficient branching method, that is, how many branches should a vessel branch into for best efficiency?

The answer is provided in Fig. 11.8.

Figure 11.8 shows that the number of branches at order number 1 is always minimum at half the total number of branches, N. It can be said that energy therefore minimizes when equal branching occurs. This is also known as dichotomous branching. Hence, the most energy-efficient branching network should have the most total number of branches and equal branching. This concept also applies to the larger vessels. Lefevre (1983) studied the input impedance of the pulmonary artery using a fractal model of the lung. Lefevre also examined the pulmonary artery model impedance over a full range of frequencies. This impedance is important because it is the dynamic load of the right heart.

This problem is not only useful to study the structural design of the pulmonary circulation but also for application to right heart failure. The final result of the analysis revealed an optimized fractal design of the lung in terms of best vessel sizes and lengths. Ultimately the optimized fractal lung was shown to conform to measure

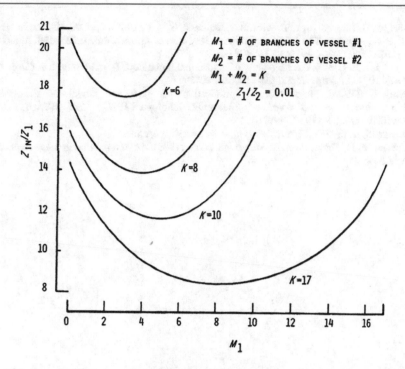

Fig. 11.8 Input impedance as a function of the number of branches at vessel order 1 for the fixed total number of branches. (Reproduced from Wiedeman et al., 1981)

pulmonary pressure and flow data of humans. The general rule so presented here is that the fractal design is most often the best design that physiology has chosen for the human cardiovascular system. It may be that the fractal design could be the best design as other biological systems are explored in this respect.

Problems

1. The power of the pulse is greatest at what location in the cardiovascular system.
2. Where is the pulsatile power the lowest in the cardiovascular system?
3. Do more arterial branches increase the power consumption of the cardiovascular system? Explain.
4. Which branching system is more protective of a blood clot – Strahler or dichotomous? Explain your answer.

References

Bassingthwaighte, J. B., King, R. B., & Roger, S. A. (1989). Fractal nature of regional myocardial blood flow heterogeneity. *Circulation Research, 65*(3), 578–590. https://doi.org/10.1161/01.res.65.3.578

Goldberger, A. L., Bhargava, V., West, B. J., & Mandell, A. J. (1985). On a mechanism of cardiac electrical stability. The fractal hypothesis. *Biophysical Journal, 48*(3), 525–528. https://doi.org/10.1016/S0006-3495(85)83808-X

Hoffman, J. I. E. (1995). Heterogeneity of myocardial blood flow. *Basic Research in Cardiology, 90*(2), 103–111. https://doi.org/10.1007/BF00789440

Lefevre, J. (1983). Teleonomical optimization of a fractal model of the pulmonary arterial bed. *Journal of Theoretical Biology, 102*(2), 225–248. https://doi.org/10.1016/0022-5193(83)90361-2

Noordergraaf, A. (1978). *Circulatory system dynamics*. Academic.

Wiedeman, M. P., Tuma, R. F., & Mayrovitz, H. N. (1981). *An introduction to microcirculation*. Academic.

Recurrence and the Embedding Dimension

12

12.1   Introduction

In the earlier chapters, it was shown that the response of nonlinear dynamic systems can be chaos. Several methods of studying chaos were introduced including the method of Fourier analysis. Unfortunately, Fourier applies to mostly linear sinusoidal systems. Hence, the interpretation of Fourier when applied to nonlinear systems is somewhat questionable. Moreover, Fourier forces the assumption of a sinusoidal model that may be restrictive. Nevertheless, Fourier is still useful keeping in mind this deficiency. With these caveats in mind, the Fourier transform of a chaotic system is reproduced here in Fig. 12.1.

Referring to Fig. 12.1, the general observation is that chaos is a broad-spectrum signal. It can also be seen that large-amplitude peaks exist at some frequencies. Usually, the frequencies at which these peaks occur are multiples of each other due to the property of period doubling that frequently occurs in nonlinear dynamics. It can also be observed that these large peak frequencies may not be present consistently. To further strengthen the observation of chaos, it may be observed that the power spectrum magnitude is inversely related to frequency. That is to say, it has the properties of Fractal time as studied in the fractal time chapter. Beyond these basic observations, Fourier is not much more useful for the nonlinear system. This leads to the question – can frequency peaks be measured without Fourier by an alternative method that is not restricted by linearity?

One method of analysis that relieves the Fourier problem is recurrence analysis (Webber & Zbilut, 1994). Recurrence analysis attempts to find periodicities with a data time series. It does not depend on linearity and is applicable to nonlinear systems. Recurrence analysis also relieves the assumption of data stationarity. Moreover, recurrence may be employed as a means of detecting nonstationarity.

Before describing recurrence, the concept of embedding space must first be defined. From the previous chapters, it is clear that fractal analysis applies best to two- or three-dimensional objects. Unfortunately, most time-series data are generally in the form of a single dimension. Embedding space allows one-dimensional

G. Drzewiecki, *Fundamentals of Chaos and Fractals for Cardiology*, https://doi.org/10.1007/978-3-030-88968-5_12

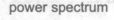

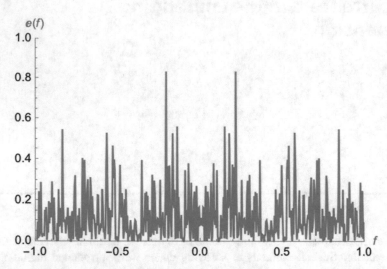

Fig. 12.1   Fourier transform of the chaotic population system example in Chap. 3. (Created from Mathematica's Wolfram Demonstration Project, Wolfram.com)

data to be converted into a multidimensional spatial object. Starting with a single-dimension time series, for example, let $S(t_i)$ = time-series data. To construct the first embedding dimension m = 1, define the array $S_{m=1} = [s(t_i), s(t_{i-1})]_{Sm=1}$ The signal S is now a two-dimensional array and the one-dimensional time series has now been reconstructed into a 2D object for which the fractal analysis methods for planar areas can be applied. Continuing on with embedding, m = 2 is then created from the first and second delay time points as an array $S_{m=2} = [S(t_i), S(t_{i-1}), s(ti_{-2})]$ that is now a three-dimensional vector or volume object to find its fractal dimension.

The embedding procedure can be repeated to create higher dimensional objects so that the mth embedding is S_m, where an m-dimensional object then is generated.

After creating the embedding structure of the time-series data, recurrence analysis identifies periodic behavior in the embedding. Once the embedding structure has been generated, the method of fractal analysis may be applied to study the embedded objects. In this chapter, it is desired to locate periodic behavior in the embedded object so that recurrence analysis is now described.

12.2   Recurrence Analysis Procedure

Use the following steps to perform a recurrence analysis:
1. Find the geometric distance between all vectors $\bar{D}_{ij}$ and their magnitude $\bar{D}_{ij}$ as defined in Eq. 12.1:

$$\vec{D}_{ij} = \left\{ S_m\left(t_i\right),\ S_m\left(t_j\right) \right\}$$

$$\vec{D}_{ij} = \sqrt{\left(S\left(t_i\right)S\left(t_j\right)\right)^2 + \left(S\left(t_{i1}\right)S\left(t_{j1}\right)\right)^2} \tag{12.1}$$

2. Search all $\vec{D}_{ij}$ to find the vector of the maximum distance, $\vec{D}_{ijMAX}$. Locate the distances that are smaller than $\vec{D}_{MAX}$ by a fraction of $F < 1$.

$$\vec{D}_{ij} < F \cdot \left.\left|\vec{D}_{ij}\right|\right|_{MAX}$$

3. Identify all vectors that satisfy condition #2.
4. Lastly, plot all points IJ identified in #3.

Step 4 is the recurrence plot of the time series. To understand its meaning, the plot is a summary of every point that indicates a close relationship between the two vectors ij and thereby suggests a periodic behavior. Note that the analysis procedure does not rely on linearity, and no specific assumption was applied to the data with all data analyzed. The value of F that is chosen need only be less than 1. Its purpose is to screen out data points that possess a weak relationship with each other. The F value may be adjusted experimentally to reject possible random relationships.

Next view an example of a recurrence plot created from the data of Fig. 12.2 obtained from a Henon attractor . This is a system of two nonlinear differential equations in x and y that produces an xy Poincare plot called the Henon attractor.

The time-series data of Fig. 12.2 was analyzed using recurrence analysis to obtain the recurrence plot of Fig. 12.3. To illustrate how the recurrence plot successfully locates periodicity in data, the original time series was randomly shuffled to destroy all time relationships. The recurrence plot was then constructed for the shuffled data. Both recurrence plots are shown for comparison in Fig. 12.3. Notice the appearance of diagonal groupings of points in Fig. 12.3a. These points indicate the

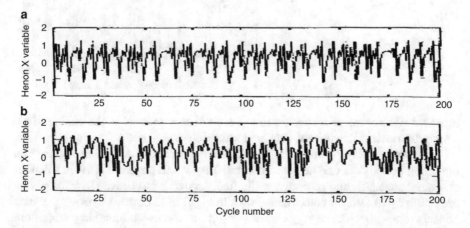

Fig. 12.2 Time solution of the Henon attractor of the dx/dt and dy/dt equations versus cycle number. (Reproduced from Webber & Zbilut, 1994)

Fig. 12.3 Recurrence
plots of Fig. 12.2 Henon
model data. Top graph is
the recurrence plot of the
raw data. Bottom graph is
the recurrence plot of the
randomized Henon data.
(Reproduced from Webber
& Zbilut, 1994)

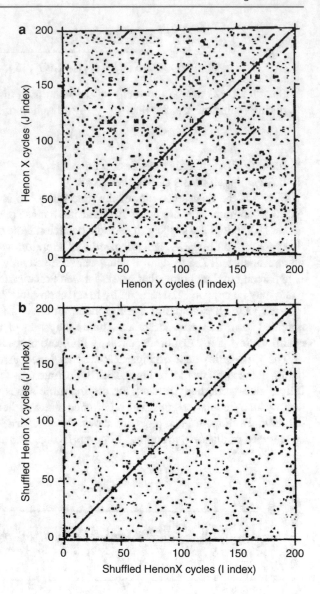

repeated occurrence of a specific time vector that is present in the chaotic data. These disappear following the data randomization in Fig. 12.3b.

Recurrence analysis was used to analyze a respiratory data time series. The breathing period data of a rat was recorded. The rat was observed to be in a resting and quiet state [A], and secondly, while freely moving and active [B] with the raw breathing cycle data for both states shown in Fig. 12.4. The raw physiological data appears chaotic with large variability even in the quiet breathing condition. Recurrence plots of the rat breathing data are shown in Fig. 12.5.

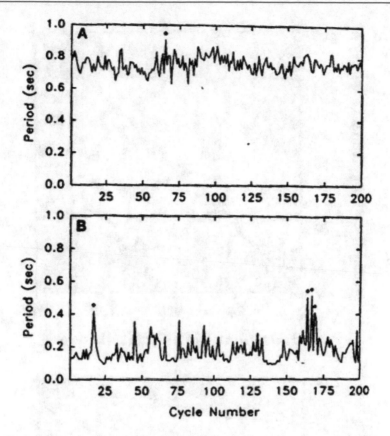

Fig. 12.4 Breathing cycle time-series data obtained from a rat for quiet breathing A and during free activity B. (Reproduced from Webber & Zbilut, 1994)

Next, review the interpretations of a recurrence plot. The first piece of information that the recurrence plot provides is that every point indicates a repeated vector of the system. This can be further interpreted as every point is the repeated state of the system. Secondly, groupings of points and visual structure indicate determinism as opposed to isolated points that are random. Single isolated points are therefore random states. The ij diagonal line is always present and represents the ij vector related to itself. Other diagonal lines indicate a separate state that is repeating itself. The recurrence plot also shows regions of the plot that contain no points. These clear areas indicate that a data transient has occurred. Referring to quiet breathing, there are many shapes of adjacent points showing a visual structure. Overall, it can be concluded that the quiet breathing data contained many repeated states. Similar structures are exhibited in the active breathing, but there are many more regions clear of points. This indicates transient activity as a consequence of the rat moving freely about. The recurrence analysis very easily handles the rat's movements by generating the clear areas of no points. On this point, recurrence analysis can handle such nonstationary data while the Fourier would have generated many erroneous

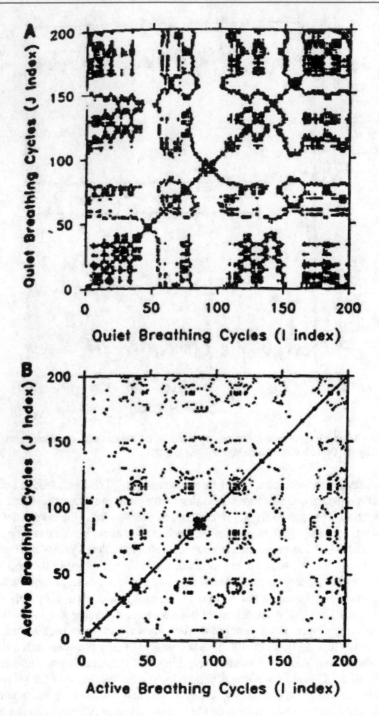

Fig. 12.5 Recurrence plots for the rat time-series data of Fig. 12.4. Plot A is for the quiet breathing and plot B is for breathing during activity. (Reproduced from Webber & Zbilut, 1994)

frequencies. Further point groupings are the vertical or horizontal groupings. These represent vectors that repeat but at different times. Lastly, the diagonal line groupings can be related to the Lyapunov exponent of the system.

In summary, this chapter has introduced recurrence analysis as an alternative means to find periodicities in data besides the Fourier analysis that may produce erroneous results in chaotic systems. The existence of structure in the recurrence plot supports the case that an unknown data set is from a deterministic system as opposed to random, as demonstrated in Fig. 12.3.

Problems

1. What frequencies have the largest magnitude in a chaotic signal?
2. Given the time series x_1, x_2, x_3, x_4, write the array for the second embedding of the signal.
3. What does a single point in a recurrence plot represent?
4. What are the benefits of recurrence analysis over frequency analysis?
5. What information does a recurrence plot provide about an unknown chaotic system?

Reference

Webber, C., & Zbilut, J. (1994). Dynamical assessment of physiological systems and states using recurrence plot strategies. *Journal of Applied Physiology (Bethesda, Md. : 1985)*, 76, 965–973. https://doi.org/10.1152/jappl.1994.76.2.965

Part III
Cardiovascular Dynamics Applications

Nonlinear Flow Dynamics and Chaos in a Flexible Vessel Model

13

13.1 Introduction

Now in Part III of this book, some applications of nonlinear dynamics will be applied to current research of the cardiovascular system. A particularly interesting problem is that of blood flow in the region of diseased vessels. Conventional fluid dynamics is difficult to apply for this kind of problem due to the fact that a blood vessel is a flexible structure. The usual fixed boundary conditions of the structure cannot be employed. This causes the vessel structure to become pressure and flow dependent. This in turn causes the fluid system parameters to become dependent.

Prior studies of flow in the region of a stenosis have generally focused on the flow velocities in the vicinity of the plaque dome narrowing of a diseased vessel (Kok et al., 2016). These researchers conclude that flow velocity becomes high in the narrowed vessel and therefore may cause vessel wall damage due to higher flow velocity shear stress. Such studies usually apply the computational fluid dynamics software and make the assumption of a rigid vessel structure. Due to the limitations of most computational fluid software, it is often assumed that flow is also nonoscillatory. In this chapter, a computational hemodynamic model will be developed that relieves the assumptions of a rigid stenotic vessel and steady pressure. This model will be further applied to calculate flow in the vessel to reveal the flow dynamics. The nonlinear dynamic analysis methods presented in the earlier chapters will be applied to examine the resulting fluid dynamics.

The pressure and flow dynamics of flexible vessels was applied to the arterial stenotic blood flow problem (Barton-Scott & Drzewiecki, 1999, 2000; Drzewiecki et al., 1997). The hemodynamics of an artery can be modeled using three elements: flow resistance, flow inertance, and volume compliance (Field & Drzewiecki, 1997).

The original version of this chapter was revised. The correction to this chapter is available at https://doi.org/10.1007/978-3-030-88968-5_16

Tracy Barton-Scott, J.K. Li, and J. Kedem also have contributed to this chapter.

© The Author(s), under exclusive license to Springer Nature Switzerland AG 2021, Corrected Publication 2022
G. Drzewiecki, *Fundamentals of Chaos and Fractals for Cardiology*, https://doi.org/10.1007/978-3-030-88968-5_13

Fig. 13.1 A noninvasive measurement of the arterial pressure–area relationship from a normal human brachial artery. (Reproduced from Drzewiecki & Pilla, 1998)

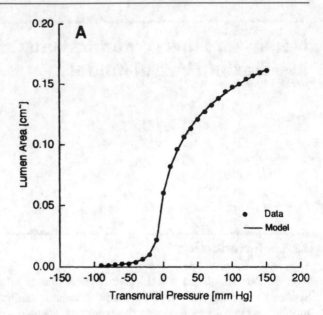

These elements are normally constant in a rigid vessel and result in a linear second-order system. Normally, this modeling approach applies well to arteries because normal physiological pressures maintain a circular lumen shape of the vessel.

Unfortunately, when the transmural pressure (inside minus outside pressures) is near zero, as in veins, the linear hemodynamics vessel model is no longer valid due to the highly nonlinear relationship between pressure and lumen area (Drzewiecki et al., 1997). For example, a medium-sized arterial pressure–area relationship is shown in Fig. 13.1.

Figure 13.1 reveals that a vascular lumen cross-sectional area is nonlinearly related to pressure in a normal human artery. Notice that the lumen area in Fig. 13.1 is nearly zero for low pressures. This means that the vessel is nearly closed and does not permit flow. This vessel property can have a dominant influence on flow. There are several normal and pathological situations where the transmural pressure of a vessel can be reduced to low or even negative levels, thereby enforcing this effect. For example, in diseased narrowed arteries, low transmural pressure can be found in vessels passing through contracting muscle under an occlusive blood pressure cuff and in the veins (Field & Drzewiecki, 1997).

The arterial pressure–area relationship can be a strong determinant of flow when it is operative. The nonlinear pressure–area relation is now applied to study blood flow dynamics in a segment of a flexible vessel with pulsatile flow now used to more accurately model an arterial dynamic system with nonlinear properties.

13.2   The Model Geometry

Two segments of an artery were placed in a series arrangement (Fig. 13.2). This corresponds with a conduit coronary vessel feeding multiple small arteries. The proximal segment is maintained at positive pressure and fully distended. The distal

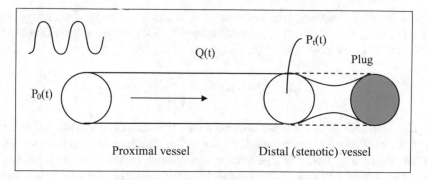

Fig. 13.2 Geometric model of the flexible vessel model

segment was subjected to positive external pressure so that the transmural pressure was reduced to zero and the vessel would collapse, thus simulating a flexible stenosis (Drzewiecki et al., 1997). A sinusoidal pressure was applied to the proximal vessel. The end of the collapsible vessel was plugged so that mean flow was forced to be zero. Therefore, only the dynamic pressure and flow characteristics were analyzed. This limitation was added to simplify the model, which can be removed if desired by changing downstream conditions. A previous arterial segment flow model contained a mean flow, and consistent results were obtained (Field, 1995). The mean transmural pressure on the collapsible segment was maintained near its buckling pressure. The transmural pressure and lumen area relationship of the canine carotid artery was previously measured and modeled (Drzewiecki et al., 1997).

13.3   Arterial Segment Dynamic Flow Model

A lumped fluid dynamic model was implemented to represent the vessel configuration. Distributed properties were not considered here. Lumped dynamic vessel impedances were used to model pressure and flow in the segment following the analysis of Noordergraaf (1978). Arterial flow resistance and flow inertance of the proximal flow segment were taken to be in series. The flow resistance, R, was obtained from the Poiseuille flow equation:

$$R = \frac{8\pi\eta l_t}{\pi^2 r^4}$$ (13.1)

where η is the fluid viscosity (0.03 Poise for canine blood at 37 °C), l_t is the proximal segment length, and r is the proximal segment radius . The flow inertance, L, was determined from the mass of the fluid within the volume of the proximal vessel assuming a cylindrical geometry:

$$L = \frac{\rho l_t}{\pi r^2}$$ (13.2)

where ρ is the fluid density of blood.

For the distal segment, the lumen area, A, is a function of the transmural pressure, $P_{t(P\ external)}$ according to $_{inside-P}$ relationship (Eq. 13.3) obtained from Drzewiecki et al. (1997) and is graphed in Fig. 13.1:

$$P_t = a\left(e^{b(A-A_b)/A_b} -1\right) - E\left(\left(\frac{A_b}{A}\right)^n -1\right) + P_b, \qquad (13.3)$$

Equation 13.3 parameters are calibrated with the values that correspond with the type of blood vessel being modeled. Where a is the elastance scale modulus, b is the elastance rate modulus, E is a parameter representing wall bending stiffness, and n is an exponential constant describing the degree of curvature of the hyperbolic relation A_b/A. A_b and P_b are the lumen area and transmural pressure at vessel buckling. The canine carotid artery was modeled. All parameter values for a canine carotid artery are provided in Table 13.1.

Equation 13.3 introduces nonlinearity into the vessel segment model where it would otherwise be represented as a constant arterial compliance when assuming linearity. Equation 13.3 then provides the arterial volume, V, from the cross-sectional area, A. Assuming a cylindrical vessel $V = l_{t}*A$, which is used in the model differential equations below. The vessel radius is then $r = \sqrt[2]{A/\neq}$. The proximal vessel is a constant radius and treated as fully distended.

Two differential equations were used to model the pressure and flow in both segments of vessel. Equation 13.4 represents flow continuity in the distal artery. Equation 13.5 is the equilibrium equation or the sum of all pressure drops due to each element.

$$\frac{dV}{dt} = Q \qquad (13.4)$$

$$\frac{dQ}{dt} = \frac{P_L}{L} = \frac{P_0 - RQ - P_t}{L} \qquad (13.5)$$

Table 13.1 Nonlinear arterial segment model initial parameters

Parameter symbol	Value	Parameter description
R	48.89	Flow resistance
η	0.03 P	Fluid viscosity for canine blood at 37 °C
l_t	40.0 cm	Proximal vessel length
r	0.5 cm	Proximal vessel radius
a	11.9	Elastance scale modulus
b	0.438	Elastance rate modulus
E	4.14	Wall bending stiffness
n	3.76	An exponential constant describing the degree of curvature of the hyperbolic relation A_b/A
A_b	0.0189	Lumen area at vessel buckling
P_b	−0.64	Transmural pressure at vessel buckling

where V is the lumen volume, Q is the fluid flow, P_L is the pressure drop across the inertance, and P_0 is the vessel driving pressure. In terms of nonlinear systems, the vessel model may be considered to be a sinusoidally driven nonlinear system.

The dimensions of the upstream vessel were adjusted to approximate a canine coronary artery. Since the damping factor determines whether unstable flow will be obtained, the damping coefficient and damping factor were determined. For the parameters employed here, the damping factor was less than or equal to 0.05, indicating that damping is minimal in this model:

$$\text{damping coefficient factor} = \alpha = \frac{R}{2L} \tag{13.6}$$

$$\text{damping factor} = \frac{a}{\omega_0} \leq 0.05 \tag{13.7}$$

The sinusoidal driving pressure is

$$P_0 = \text{MAP} + A_0 \sin(\omega t) \tag{13.8}$$

where P_0 is the input pressure, MAP is the mean arterial pressure, and A_0 is the peak pressure amplitude.

13.4 Computational Flow Evaluation

Due to the nonlinearity introduced by Eq. 13.3, the model differential equations were solved numerically using the Runge–Kutta algorithm. The system model was first tested by using some basic linear systems methods. The frequency response of the model was obtained by performing a frequency sweep of the driving pressure. The peak amplitude of the driving pressure was maintained at less than 10 mmHg so that the nonlinearity of the distal vessel was minimal.

The vessel was then analyzed under a constant frequency while changing the sinusoidal driving pressure. The mean transmural pressure of the collapsible segment was maintained at zero. Thus, the vessel periodically opened and closed. The driving pressure frequency was set approximately equal to the linear resonant frequency. The flow response was then computed for increasing peak amplitudes of the driving pressure, A_0.

Using the methods of nonlinear analysis methods described in Chaps. 2 and 3, the flow response was evaluated to test for the chaos, and a number of graphs were developed. First, phase (Poincare) plots were created to compare the input pressure to the distal vessel pressure. Second, the flow waveform was also computed to examine for chaotic flow. Lastly, bifurcation diagrams were produced to explore various steady-state solutions of chaotic flow as described in Chap. 3 and to explore the parameter space over a wide range of model parameters. To produce a bifurcation diagram, the model is run repeatedly with various model parameters. Each time the final flow values were retained and plotted against the parameter under investigation.

13.5   Nonlinear Vessel Segment Analysis

First, the frequency response of the vessel segments was obtained for small driving pressures to maintain an approximately linear system. In this case, simple second-order linear resonance results. Linear systems methods then apply. Here, the flow waveforms are sinusoidal and only vary in amplitude and phase. For example, Fig. 13.3 demonstrates the following:

– A sweep frequency response is shown to reveal a second-order resonant response.
– The vessel could be "tuned" by varying the mean arterial pressure to a different level.
– Transition to a nonlinear response by increasing the driving pressure amplitude.

The resonant frequency of the canine carotid artery varied from 2.88 Hz for $P_t = 0.0$ mmHg to 5.23 Hz for $P_t = -11.0$ mmHg.

The flow response was computed for increasing amplitudes of the driving pressure, A_0. As the pressure amplitude increased, the resulting flow was found to possess a nonsinusoidal waveform. Additionally, for amplitudes of 15 mmHg or greater, chaotic activity was evident from the aperiodic form of the resulting waves in Fig. 13.4.

Next, the linear dynamics is examined further using the Poincare plot method.

Figure 13.5 shows the phase plots that were produced using a mean arterial pressure of 5.0 mmHg and peak pressure amplitude of 1.0 mmHg.

The combination of this mean arterial pressure and the small pressure amplitude was used to ensure a positive transmural pressure and near-linear conditions. Since the transmural pressure was kept positive, the vessel did not approach closure and no chaotic behavior was noted. Figure 13.5a shows the inverse relationship between driving and downstream pressure, indicating a 180° phase shift. The circular shape

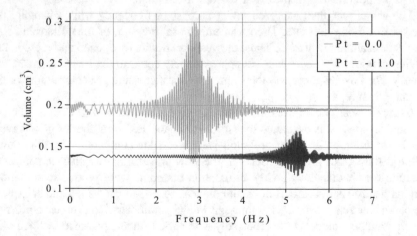

Fig. 13.3 Frequency sweep for collapsible vessel model at two different transmural pressure values

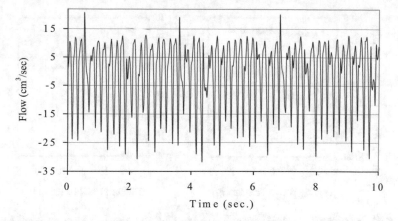

Fig. 13.4 Chaotic flow response to a sinusoidal driving pressure with an amplitude of 20 mmHg and $P_t = -11$ mmHg with the driving pressure frequency set at 5.238 Hz, approximately equal to the linear resonant frequency

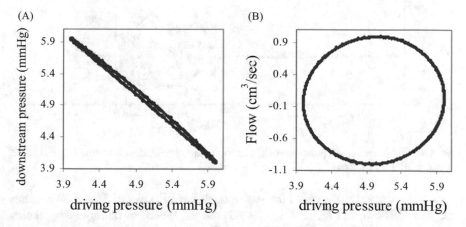

Fig. 13.5 Phase plots for a MAP of 5.0 mmHg and an A_0 of 1.0 mmHg. The case of a near-linear vessel segment. (**a**) shows the driving pressure versus distal pressure and (**b**) shows the driving pressure versus flow

of Fig. 13.5b represents a simple 90° phase shift between flow and driving pressure. Now, the Poincare plot is used to examine the chaotic flow condition of Fig. 13.4.

Figure 13.6 shows the phase plots for a mean arterial pressure of 0 mmHg and peak pressure amplitude of 20 mmHg, the same parameters used for the chaotic flow waveform produced (Fig. 13.4). The driving pressure versus downstream pressure waveform no longer shows a linear relationship. It dips dramatically once the downstream pressure approaches 0 mmHg. At this point, the plot also begins to thicken as the data is "filling" in the graph. The driving pressure versus flow waveform also shows "filling" at lower pressures. "Curve filling" is a characteristic of chaotic activity.

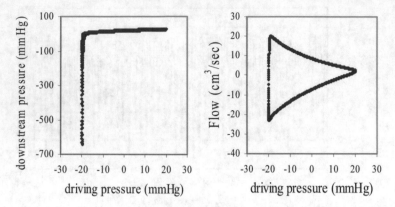

Fig. 13.6 Phase plots for a MAP of 0 mmHg and an A_0 of 20.0 mmHg. Left panel shows the driving pressure versus downstream pressure, and right panel shows the driving pressure versus flow

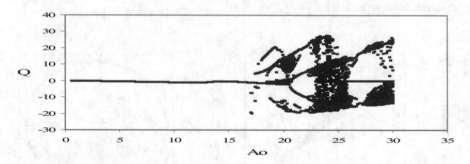

Fig. 13.7 A bifurcation diagram created with peak pressure amplitudes of 0–30 mmHg and initial flow from 0 to 40 cm³/sec

Turn now to study the vessel flow using the bifurcation plot of flow. A bifurcation diagram is shown in Fig. 13.7 and reveals the emergence of multiple frequencies and chaos for the chaotic flow condition of Fig. 13.4.

The multiple frequencies are shown when the graph bifurcates and the chaos is evident in the areas after the bifurcation that are filled in toward the right side of the plot.

Although it has so far been shown that chaos exists in this model of blood vessel flow, the blood vessel in our model reacts instantaneously. This cannot happen in reality due to viscoelastic properties that have not yet been considered. The viscous property may be increased by increasing the model damping factor Eq. 13.7 that increases the flow resistance element in the model. Since both the resistance and the inertance were obtained using the dimensions of the vessel (Eqs. 13.1 and 13.2), the total amount of the resistance was increased by a factor so that the vessel's dimensions were not disturbed.

New bifurcation plots were then constructed to determine if there was any effect on chaotic activity. Figure 13.8 shows bifurcation plots for two and three times the resistance values.

This increased the damping factor from 0.05 to 0.10 and 0.15, respectively. The bifurcation diagrams in Fig. 13.8 show a decrease in the amount of chaos, but it was not totally eliminated and there were still multiple bifurcations. To demonstrate the effect of multiple frequencies, a flow waveform is shown in Fig. 13.9. The peak pressure amplitude where multiple frequencies exist, on the bifurcation diagram produced using three times the resistance, was used.

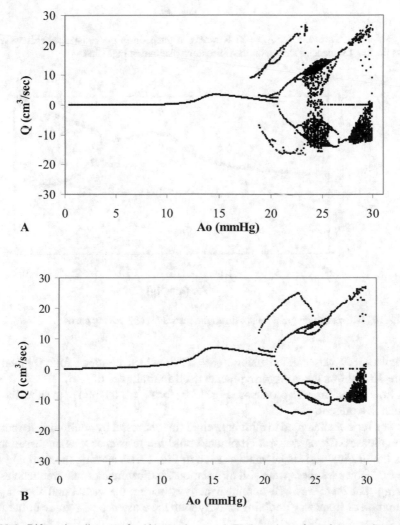

Fig. 13.8 Bifurcation diagrams for (**A**) two times and (**B**) three times the resistance of the vessel used in the production of the bifurcation diagram of Fig. 13

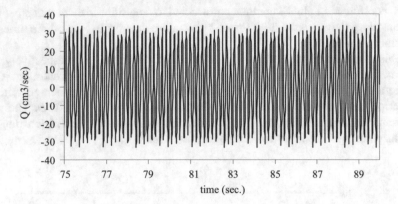

Fig. 13.9 Flow waveform for $A_0 = 22.5$ mmHg at three times the resistance. This waveform shows the four frequencies shown on the bifurcation diagram in Fig. 13.8b

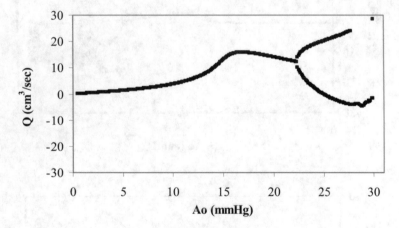

Fig. 13.10 Bifurcation diagram for six times the resistance obliterating chaos

Figure 13.9 was created using $A_0 = 22.5$, which contains four frequencies. Figure 13.9 shows the waveform repeats itself in multiples of four.

A flow waveform was produced (Fig. 13.9) for $A_O = 22.5$ mmHg to show the four frequencies present.

It was further attempted to eliminate chaos in this vessel by continuing to increase viscous losses. Chaos was not eliminated until the resistance was increased to six times the original resistance value as seen in Fig. 13.10, causing the damping ratio to be 0.3. However, there were still bifurcations, indicating multiple frequencies and showing that there was still nonlinear activity within the vessel and showing the elimination of frequency doubling activity with the existence of a single bifurcation.

13.6 Summary

Several phenomena were observed in the flexible arterial segment. First, the vessel resonance was tunable. This differs from a normal linear blood vessel that will resonate at a single frequency and no chaotic flow is present.

The vessel also demonstrated a chaotic flow response for large-amplitude pressures. This response was eliminated when the pressure amplitude was reduced sufficiently. This was shown by the phase plots and Poincare plots, which showed chaotic activity for a MAP = 0 mmHg and A_0 = 20 mmHg. These two adjustments move the function of the vessel into its nonlinear range of function. Then, the vessel acts like a nonlinear sinusoidal-driven system. The expected results are bifurcations and chaos, as shown in this chapter. The system was studied here using the tools presented in this book. This demonstrates how the methods provided in this text are useful in analyzing a nonlinear flow model. Imagine if the researcher uncovered the flow patterns that result from the vessel system. Without these tools, the researcher might be convinced that their solution was erroneous. The cardiovascular dynamic researcher can now understand how to recognize and deal with an apparently random result that is in fact real. The arterial segment model presented here is an established arterial segment model in its linear form where the distal segment is a constant compliance. The linear compliance was substituted for a flexible vessel segment. First, this introduced the nonlinearity necessary to cause chaotic flow. Second, it alerts the researcher to be aware that the presence of a flexible and collapsible vessel may lead to chaotic blood flow. This situation is not unusual in cardiovascular physiology. The diseased vessel is known to produce sound. These sounds are typically thought to represent blood turbulence, but as shown here, the flow bifurcations to chaos may lead to audible flows as well. Additionally, arteries and veins often collapse when located within contracting muscle tissues. Vessels located within the upper lobes of the lungs may collapse as they are subject to high hydrostatic pressure. Lastly, the artery under an occlusive arm cuff is subject to high external pressure leading to collapse and sounds that are audible during blood pressure determination known as the Korotkoff sounds (Drzewiecki et al., 1989).

References

Barton-Scott, T., & Drzewiecki, G. (1999). Variable resonance and Chaos in collapsible blood vessels. In M. Nowak, R. Adrezin, & D. Leone (Eds.), *Proceedings of the 25th Northeast Bioengineering Conference* (pp. 94–95). University of Hartford: IEEE Press.

Barton-Scott, T., & Drzewiecki, G. (2000). Effect of Visco-elasticity on Chaos in Collapsible Blood Vessels. In J. Enderle & L. Macfarlane (Eds.), *Proceedings of the 26th Northeast Bioengineering Conference* (pp. 31–32). University of Connecticut, Storrs: IEEE Press.

Drzewiecki, G., Field, S., Moubarak, I., & Li, J. K.-J. (1997). Vessel growth and collapsible pressure-area relationship. *American Physiological Society, 273*, H2030–H2043.

Field, S. (1995). *Nonlinear dynamic model of an arterial stenosis (atherosclerosis) and chaotic pressure and flow dynamics using a flexible-collapsible blood vessel with and external compliance load.* Ph.D. dissertation, Rutgers University and UMDNJ-Robert Wood Johnson Medical School.

Field, S., & Drzewiecki, G. (1997). Dynamic response of the collapsible blood vessel. In G. Drzewiecki & J. K.-J. Li (Eds.), *Analysis and assessment of cardiovascular function* (pp. 277–296). Springer-Verlag.

Drzewiecki, G., & Pilla, J. J. (1998). Noninvasive measurement of the human brachial artery pressure-area relation in collapse and hypertension. *Annals of Biomedical Engineering, 26*(6), 965–974. https://doi.org/10.1114/1.130

Drzewiecki, G. M., Melbin, J., & Noordergraaf, A. (1989). The Korotkoff sound. *Annals of Biomedical Engineering, 17*(4), 325–359. https://doi.org/10.1007/BF02368055

Kok, A. M., Speelman, L., Virmani, R., van der Steen, A. F., Gijsen, F. J., & Wentzel, J. J. (2016). Peak cap stress calculations in coronary atherosclerotic plaques with an incomplete necrotic core geometry. *Biomedical Engineering Online, 15*(1), 48. https://doi.org/10.1186/s12938-016-0162-5

Noordergraaf, A. (1978). *Circulatory system dynamics*. Academic Press.

The Origin of Heart Rate Variation Using a Nonlinear Modeling Approach

14

14.1 Introduction

In this chapter, we will apply the background obtained on nonlinear oscillations and bio- oscillators to the problem of heart rate variation. The heart rate is typically measured as a time series of beat-to-beat heart rate. The heart rate is typically obtained from a subject's ECG. The p or R-waves are chosen as reference waves, and the time between the occurrence of the R-wave of each heartbeat is measured. This is the period of a single beat. Period is then inverted to yield the heart rate for that individual beat. The process is then repeated for all beats sequentially. The heat rate may then be plotted as heart rate versus beat time or beat number. Sample heart rate data is shown in Fig. 14.1.

Referring to the data in Fig. 14.1, the most apparent difference between subjects is that the range and variability of heart rate are not consistent. In particular, patient B shows a heart rate graph that is normal. The other graphs are for patients of cardiac dysfunction. The heart rate plot is not clear to read and diagnose from. At first, it appears that the variability of the rate series carries useful diagnostic information. Realizing that it is the change in heart rate that may carry useful diagnostic information, various measures of heart rate variation are available. These measures analyze the heart rate time series as in Fig. 14.1. Several heart rate variability measures are summarized by Teich et al. (2000) . Perhaps the most common is the mean NN, where N is the heat rate of the Nth heartbeat. The difference between the current and next beat heart rate is N-N or NN. The arithmetic average of several beats of NN is then the mean NN. While these are some useful measures of heart rate variation, a clear origin of heart rate variability (HRV) is not yet available. Basic physiology implicates the baroreceptor reflex mechanism as a possible source, but the details of the variability are not well defined. This makes it difficult to relate HRV to any cardiovascular dysfunction. Since consumer devices are readily available to record subject HRV, a dynamic modeling of HRV would be useful to interpret these data.

A. Narakornpichit also have contributed to this chapter.

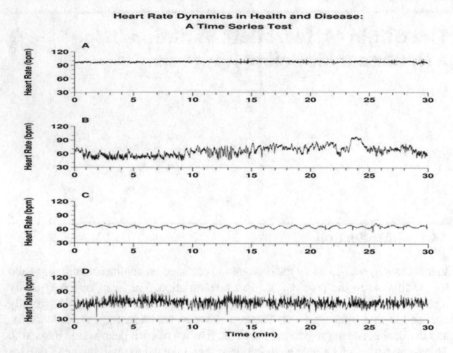

Fig. 14.1 The heart rate time series captured from four different patients. (Reproduced from Goldberger et al., 2002). Panel **B** is a normal subject. Panels **A**, **C**, and **D** are from patients with heart dysfunction

HRV falls into the topics introduced earlier in Chap. 6. For example, the heart rate time series has been studied as a Fractal time series as well as applying other nonlinear dynamic analysis, such as Poincare plots and spectrum analysis. In this chapter, a nonlinear dynamic approach to HRV will be applied instead.

Some researchers have applied purely mathematical approaches to analyze the heart rate data such as Teich et al. (2000). While this kind of approach has been successful in modeling the time series, it is not capable of relating to the physiological mechanism that caused the heart rate data.

14.2   Physiological Modeling Approach to HRV

Another approach to modeling the heart rate has been to develop systems representation of the cardiovascular system in its entirety. This approach was first employed by Guyton [REF]. Such approaches are closely tied to the function of the various organ systems involved. But the end result model is a system of linear differential equations that unfortunately do not generate the heart rate variations normal to humans. Ursino and Magosso (2003) applied the cardiovascular modeling approach with the addition of a vascular bed myogenic stochastic variability to finally generate heart rate variability.

Some approaches to heart rate are purely mathematical and can successfully model a time series but do not relate to a physiological mechanism. These approaches incorrectly assume a stochastic process.

Other more complete cardiovascular system models do not reveal the cause of HRV itself. Likely these approaches have been inadequate due to the assumption of linearity. In this chapter, it will be shown how to apply the knowledge gained from the prior chapters and will focus on modeling the key rate producing physiology that causes HRV. As such, the modeling methods applied here will not ignore the nonlinear properties of the rate-generating system as have been simplified in the past.

14.3   Heart Rate Regulation Physiology: The Baroreceptor Reflex System

To begin our analysis, it is first useful to focus on the physiological system that regulates the heart rate, that is, the baroreceptor reflex arc. A systems diagram of the baroreceptor system is provided in Fig. 14.2.

The baroreceptor system shown here is a complete representation for the reflex regulation of blood pressure by the baroreceptor. It is a negative feedback loop. Of great interest is the frequency produced by the SA node, which is the pacemaker of the heart.

We contend that, if this system is modeled correctly for the physiology and nonlinear elements, then the HRV should emerge. We will begin by going around this loop and modeling each element accurately. MATLAB and SIMULINK, together with the math functions for each element, are appropriate approaches to the final solution of the system. Being a loop, the beginning can be chosen at any point. Begin with the SA node.

Fig. 14.2 The cardiovascular baroreceptor system

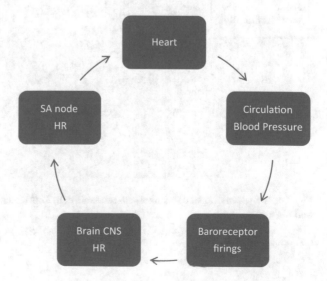

14.4   The SA Node Oscillator

The SA node is a group of cells that are self-excitable. Their frequency is modulated by the central nervous system (CNS). A nonlinear self-excitable oscillator with adjustable frequency is required. Referring back to Chap. 2, the integrate and fire oscillator is a good model for a self-excitable cell, also known as the relaxation oscillator. It will be adopted to serve as both the SA and AV node oscillators. Maintaining a system modeling format, begin with the systems model for a single self-excitable cell in Fig. 14.3.

This model follows the "integrate and fire' concept of Chap. 2. A single oscillator is shown in Fig. 14.3. It will be used to model the SA and AV nodes within the baroreceptor model. The left-side input to the oscillator will be the signal that directs the firing frequency coming from the baroreceptor. The integrator is the 1/s block. The integrator outputs to a logic block that "fires" the action potential when a threshold level is reached. The input signal also serves as a constant of integration. It also determines the frequency of oscillation where frequency is directly proportional to the input value. The oscillator model is nonlinear due to the logic block, which also limits the output magnitude of the oscillator. Hence, it may be defined to be a limit cycle oscillator. Continuing to the next element of the model, the heart.

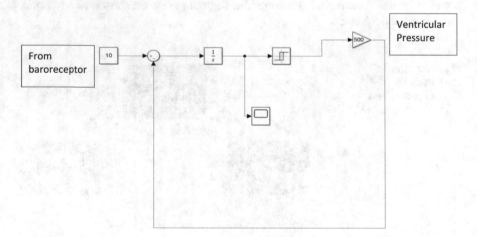

Fig. 14.3 Systems model of a single self-excitable cell. This model is duplicated for both the SA and AV nodes

14.5   The Ventricular Pressure

For simplicity, the heart will be assumed to be a pressure source. The logic block will also serve as the model for periodic ventricular pressure. The high limit sets the systolic pressure, and the low limit sets the end diastolic pressure. The output of this oscillator was provided earlier in Chap. 2.

14.6   The Arterial Circulation and Blood Pressure

The next block is the systemic arterial circulation. For this, the aortic Windkessel model was chosen (Noordergraaf, 1978). Included in the arterial model will also be the aortic valve so that flow is only into the arterial system. The arterial system model is provided for a single differential equation in Eq. 14.1.

$$\frac{dp}{dt} = \left(-p/Ra + Pv/Ro\right)/Cs \tag{14.1}$$

The arterial blood pressure is p, and Pv is the ventricular pressure. Ra is the arterial flow resistance, Cs is the arterial compliance, and Ro is the aortic valve flow resistance. For $Pv > Ps$: $Ro = Ro =$ open valve flow resistance. For $Pv < Ps$: $Ro = \infty =$ valve closed resistance.

14.7   The Baroreceptor Function

The baroreceptor is a node of stretch-sensitive cells that are located on the carotid artery. These cells sense the stretch of the carotid artery as a measure of blood pressure. The baroreceptor is also self-excitatory and will fire action potentials at a frequency that is proportional to the blood pressure. The baroreceptor response is also nonlinear. It will be modeled here empirically using a sigmoid function [REF]. The baroreceptor function is shown in Fig. 14.4 along with its plot. The input variable is μ, the arterial blood pressure. The function outputs the firing rate signal, which has been normalized and is rescaled to the proper range for the SA node oscillator. The 90 parameter sets the average blood pressure to equal 90 mmHg for a human. The five parameter adjusts the sensitivity of the baroreceptor. It can be adjusted experimentally for best pressure regulation.

14.8   CNS Function

The next model block is the CNS block. The CNS serves to properly adjust the heart rate by modulating the SA node. It also divides the regulation into the sympathetic and parasympathetic neural paths. In our model, the 0.5 firing rate is chosen as the mean firing rate. So, baroreceptor rates above 0.5 are sympathetic and increase the heart rate. Values below 0.5 are parasympathetic and slow down the heart rate. In the model, this appears as a simple addition function as the SA heart rate (baro firing rate 0.5). The CNS function provides some additional functions. First, it ensures that a negative feedback reflex pathway is maintained as required for homeostasis of blood pressure. Secondly, it fixes the SA oscillator at a minimum heart rate. This prevents the heart rate from going to zero frequency or cardiac arrest. Physiologically, both pacemaker oscillators possess this characteristic. That is, they will continue to operate even with no input. This completes all elements of the heart rate model. Next, all blocks are connected into the complete system representation shown in Fig. 14.5. Notice the addition of the AV node since the SA node electrically drives the AV node physiologically. All block functions have been described separately above. The model shown was directly loaded into MATLAB–SIMULINK with the appropriate functions inserted into each block described in the preceding sections.

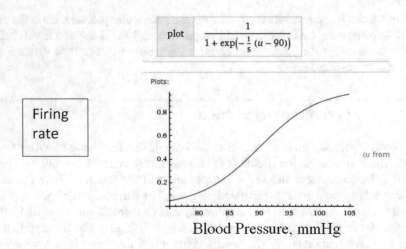

Fig. 14.4 The empirical sigmoid function used to model the baroreceptor function and a sample plot

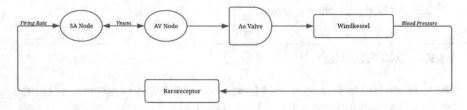

Fig. 14.5 Systems representation of the heart rate regulation model

14.9   Modeling Heart Rate and Variability

At the start of this modeling effort, the assumption was that a properly modeled baroreceptor reflex arc should automatically possess arrhythmia including variability. With model elements inserted and properly calibrated, the model outputs were computed as the SA node, AV node, and blood pressure.

The SA node and AV nodes are shown in Fig. 14.6. This plot was obtained for steady-state regulation conditions. No variability is present. It represents a basic test of the physiology. For example, the SA and AV are in frequency and phase lock. Normally physiology would require that both oscillators operate at the same frequency. Note that, had we chosen in a linear fixed oscillator, this basic physiology would not occur. Instead, the nonlinear biological oscillators modeled here operate consistent with the physiology.

A phase plot was created in Fig. 14.7 for the steady-state condition. A simple limit cycle at a single frequency is seen.

Next, the heart maximum pumping pressure was adjusted experimentally to observe arrhythmias. Figure 14.8 shows an arrhythmia.

A partial phase lock is evident along with a 1:2 frequency where the SA oscillator operates twice that of the AV. Cardiologists define this case as AV node block. The heart pumps too slowly in this condition. Continuing with varying the heart pump pressure, more arrhythmias can be found. For a normal cardiac pressure, the model demonstrated a normal sinus arrhythmia. Normal human heart rate variability is shown in Fig. 14.9.

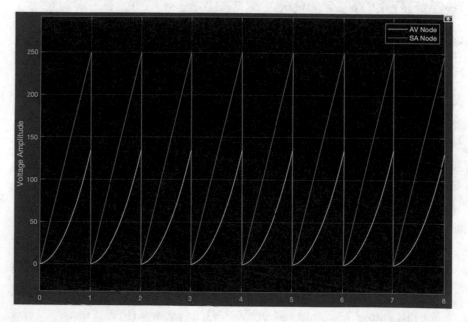

Fig. 14.6 The SA and AV node outputs for steady-state conditions of the model Some nonlinear methods can be applied now

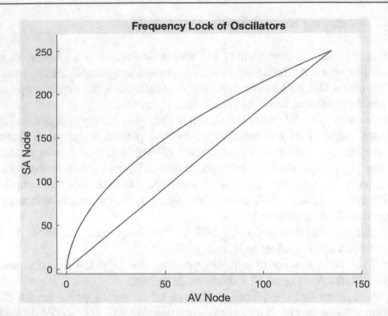

Fig. 14.7 Phase plot of the SA and AV oscillators against each other for phase lock

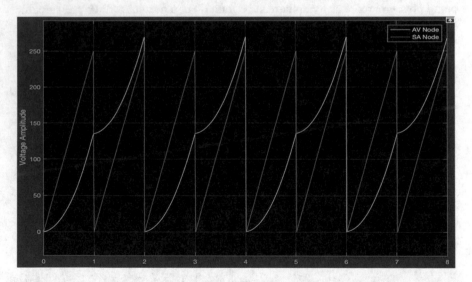

Fig. 14.8 The SA and AV node oscillators for an arrhythmia. The rhythm shown is a 1:2 rhythm

Notice that the heart period will change every beat. Now, the standard heart rate variability measures can be applied to the model as for human data. For example, the change in beat-to-beat period, NN, of the model is presented in Fig. 14.10.

For this normal heart rate variation (HRV), an average HRV of mean NN was 132.55 mSec. The range of HRV corresponds with that measured on normal human

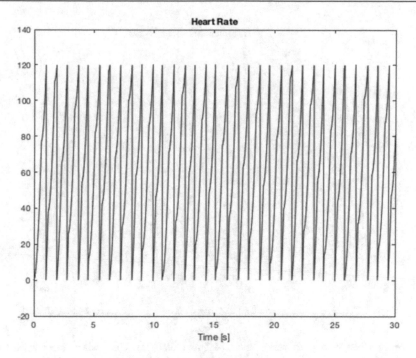

Fig. 14.9 Heart rate variability for a normal sinus arrhythmia

subjects (Teich et al., 2000). Referring to Fig. 14.10, the model also shows the expected increase (sympathetic activity) and decrease (parasympathetic activity) that is typical of human HRV. These basic tests confirm that the model developed here is representative of human HRV behavior, as shown in Chap. 6.

The HRV time series generated by the model may be examined for its fractal properties. A Fourier transform was applied to the model HRV data to create the Fourier frequency spectrum shown in Fig. 14.11.

The model spectrum is characteristically human as measured by Goldberger et al. (2002). Moreover, it exhibits fractal property. Observe that the magnitude is inversely related to the frequency (see Chap. 6). The dominant frequencies present in the HRV spectrum are of low frequency, less the 1 Hz. This has also been observed by physiologists who attribute these frequencies to the CNS response. Of course, since this physiology was included in our model, they do appear in its spectrum as well. The primary difference between these model results and others is a missing frequency near 0.1 Hz. This is the frequency of breathing (Ursino & Magosso, 2003). Since respiration was not included in the model here, no such frequency peak exists.

Lastly, the model was used to explore the HRV while altering the strength of the heart. The strength of the heart is defined as contractility. Contractility corresponds with maximum heart pressure in the model. The mean HRV was graphed with respect to contractility in Fig. 14.12.

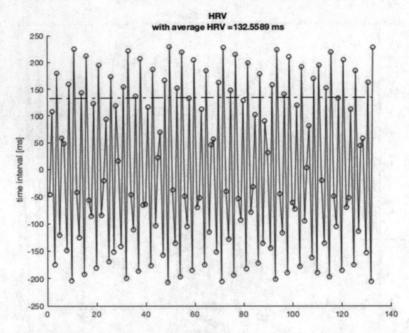

Fig. 14.10 Change in heart period from beat to beat (NN). The dashed line indicates the mean HRV

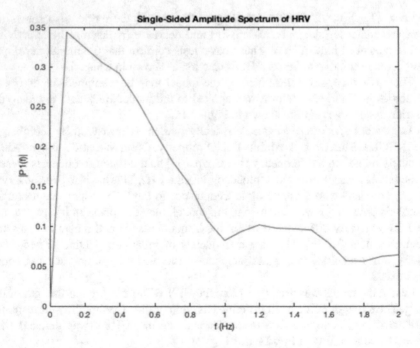

Fig. 14.11 The Fourier spectrum of the model HRV time series

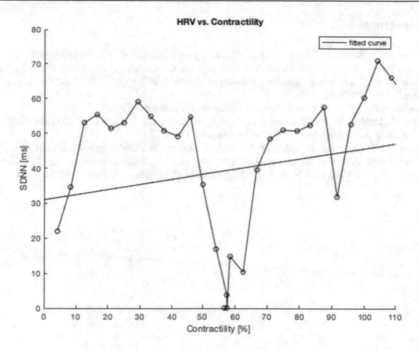

Fig. 14.12 HRV versus heart contractility. The line is linear fit to the data

A clear inverse relationship exists between HRV and contractility. This graph reveals that a healthy and strong heart is associated with greater HRV. This has also been observed in human patient data as shown in the beginning of this chapter.

14.10 Summary and Conclusion

Summarizing what has been discussed in this chapter, the problem of heart rate variability (HRV) was highlighted as another application of the content of this book. Others have attempted to solve this problem using physiological modeling but were not successful primarily due to the limitations of a linear approach. Also, fundamental to the resolution of the problem here was the application of a true biological oscillator. A second aspect that contributed to the ability to solve this problem was the realistic modeling of the physiological reflex arc without the linearity restrictions. All nonlinear functions were treated as they appear in human physiology. These two important items led to a resolution of the problem. Historically, the origin of HRV may have been missed simply because it is tempting to linearize problems. This chapter is a fitting end to all that has been covered in this book and hopefully demonstrated to the reader that nonlinear problem-solving should not be simplified but treated as a whole part of the solution. Moreover, these last application chapters have emphasized that nonlinear function often appears in physiology. The tools provided in this book have provided the reader with a way of recognizing nonlinear behavior and how to characterize it.

References

Goldberger, A. L., Amaral, L. A., Hausdorff, J. M., Ivanov, P., Peng, C. K., & Stanley, H. E. (2002). Fractal dynamics in physiology: Alterations with disease and aging. *Proceedings of the National Academy of Sciences of the United States of America, 99*(Suppl 1), 2466–2472. https://doi.org/10.1073/pnas.012579499

Noordergraaf, A. (1978). *Circulatory system dynamics*. Academic Press.

Teich, M., Lowen, S., Jost, B., Vibe-Rheymer, K., & Heneghan, C. (2000). Heart rate variability: Measures and models. *arXiv: Biological Physics*, 159–213.

Ursino, M., & Magosso, E. (2003). Role of short-term cardiovascular regulation in heart period variability: A modeling study. *American Journal of Physiology-Heart and Circulatory Physiology, 284*(4), H1479–H1493. https://doi.org/10.1152/ajpheart.00850.2002

Nonlinear Oscillator Circuit Model Lab Exercise (Properties of Self-Excitable Cells)

15

15.1 Introduction: Self-Excitable Oscillations in Excitable Cells

In this lab experiment, you will use an electronic relaxation oscillator to model the timing and function of a self-excitable cell and then to observe the complex rhythms produced by periodic stimulation of the oscillator.

Some human cells possess a membrane current that prevents the membrane voltage from returning to its resting potential. Instead, the membrane potential continuously moves the membrane voltage toward threshold potential and causes an action potential to fire repeatedly at some frequency. These are self-excitable cells that produce and generate a periodic action potential. For example, the pacemaker of the sinoatrial node of the heart is such a cell. Some smooth muscle cells have this same function, specifically in the gastrointestinal tract. As long as the membrane current is present and the cell membrane is healthy, the cells will oscillate continuously. The Hodgkin–Huxley H-H model (Hodgkin & Huxley, 1952) of the nerve action potential may be modified to demonstrate pacemaker cell function. This is accomplished by adding an additional membrane current to the model (Fig.15.1) shown as current due to E_L in the figure. The remaining circuit elements are as originally defined by Hodgkin and Huxley. This additional membrane current is defined as the funny membrane current (DiFrancesco, 2010). There are two phases of pacemaker potential function. First, following an action potential, the funny current causes the membrane capacitance to charge with in direct proportion to the current I (Integrate). Second, the membrane voltage reaches the threshold potential. This opens the sodium channel, G_{Na}, to open and begin a normal action potential (Fire). Following the action potential, the potassium conductance, G_K, restores the membrane potential toward resting levels. These phases become a sequential integrate and fire process that repeats forever as long as there is voltage available. The action potential oscillations for a single Hodgkin–Huxley model of a Purkinje cardiac cell were computationally simulated and are shown in Fig. 15.2. Looking closely at the action potentials in Fig. 15.2, the portion of the voltage that begins rising from zero is due

© The Author(s), under exclusive license to Springer Nature Switzerland AG 2021
G. Drzewiecki, *Fundamentals of Chaos and Fractals for Cardiology*,
https://doi.org/10.1007/978-3-030-88968-5_15

Fig. 15.1 H-H model modified to include the funny current in the G_L branch to cause oscillation

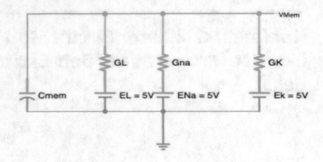

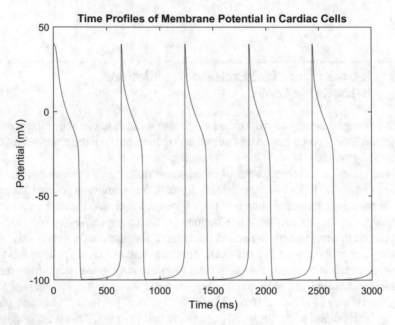

Fig. 15.2 Oscillating action potential function produced by the circuit model of Fig. 15.1 with parameters adjusted to represent a cardiac Purkinje cell

to the funny current and causes an action potential to occur once the threshold voltage is reached.

Bio-oscillator Timing Characteristics

1. **Permanent phase shift:** A bio-oscillator cell can be recognized as distinctly different from other oscillator types as it has the property of permanent phase shift. That is, if the oscillator is perturbed by say an external pulse, then the phase shift caused by that perturbance will remain forever.

2. **Phase lock:** In the case that a bio-oscillator is perturbed periodically by another oscillator. For a strong perturbance, the bio-oscillator will take on the new frequency of the perturbance. This is called phase lock or entrainment when the two oscillators start to oscillate at the same frequency.

3. **Chaos:** in a special case of phase lock, the two oscillators may take on multiples of the two frequencies. For example, there can be 1:1 lock, 1:2 lock, 1:3 lock, and so on. This situation is very similar to the cardiac electrophysiology representing the SA node stimulation of the AV node. In the case where the lock is not strong, the bio-oscillator will generate a chaos signal. That is, it will generate near noise or no apparent frequency at all.

Integrate and Fire Oscillator Circuit Model

Now, understanding the actual cell mechanism of oscillation, it is possible to engineer this kind of process into an electronic circuit. These circuits come in various forms and fall into the general category of relaxation oscillators, but timing is similar to that of the cell and is formed by an RC time constant. The most common electronic oscillator that follows this principle is the 555 timer circuit that will be used in this experiment. It follows the process of integrate and fire to cause oscillations and therefore models the timing function of a bio-oscillator fairly well. It should be noted that this experiment is not unique to the 555 IC. Other electronic relaxation oscillators may also be used. The 555 IC was chosen here since it is convenient and widely available.

Experiment Goals

In this experiment, the 555-timer IC will be used as a model of self-excitable cell oscillations. You will construct the circuit, verify its function, then periodically perturb the oscillator to observe its response to external stimulus. Lastly, you will observe the oscillator functions of frequency lock and chaos.

Equipment List

1. Sinusoidal frequency generator
2. Lab bench oscilloscope
3. Oscilloscope probe
4. DC power source (0–10 Volts)
5. Circuit breadboard
6. Supply of test board connection wires
7. 555 IC
8. 10 kohm and 500 kohm resistors
9. Two -1 micro-F capacitors
10. 10 nF capacitor

Procedures
Part 1: Basic Relaxation Oscillator Function
Step 1. Construct and Verify the Circuit Model

In this trial, you will assemble a 555 IC oscillator circuit and verify that it is functioning as an integrate and fire relaxation oscillator. Construct the circuit shown in in Fig. 15.3 on a circuit test board.

Step 2. Connect the lab bench DC source to the +Vcc point. Turn on the bench DC voltage source to 5 volts.

Step 3. The oscillator should begin to oscillate immediately with power applied. Verify that it is oscillating by connecting your oscilloscope scope probe to the output pin 3. You should see a square wave at this point. Measure the output waveform on your oscilloscope. Measure the MAX voltage MIN voltage (peak-to-peak) voltage and the frequency of oscillation. Photo your scope screen.

Step 4. Change your power source voltage to 9 volts and repeat step 2 above.

Step 5. Photo your waveform and note specific changes. Does frequency change? Why or why not?

Step 6. Move your scope probe to measure the waveform on pin 2 across the capacitor C. It should appear as in Fig. 15.4. Photo this waveform and observe that it is like cellular bio-oscillator membrane voltage in Fig. 15.2 in that it exhibits the integrate and fire phases.

Part 2. Periodic Stimulus Response of the Oscillator

Goals: In this part, you will observe that the oscillator model circuit exhibits the characteristics of phase lock, entrainment, and chaotic oscillations.

Step 1 Modify the oscillator circuit that you constructed above by removing the 10 nF capacitor from the ground and instead connect it to the lab bench sinewave oscillator as shown in Fig. 15.5. Notice that the basic oscillator circuit is unchanged except for the input of the 1.5 Hz sinewave source. The voltage at the CTL pin determines the trigger level at which the oscillator begins to discharge the capacitor. The 1.5 Hz source then periodically affects the timing of the

Fig. 15.3 (Left) Basic oscillator experimental setup and Ne555 oscillator circuit. Use element values shown on the diagram. (Right) Pin connection diagram for the 555 IC

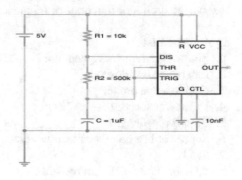

(TOP VIEW)

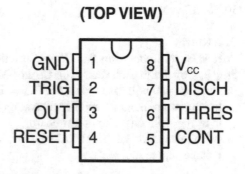

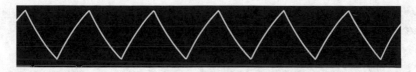

Fig. 15.4 A sample capacitor voltage wave form for the 555 oscillator circuit. A charge–discharge cycle is evident with a frequency of 1.4 Hz.

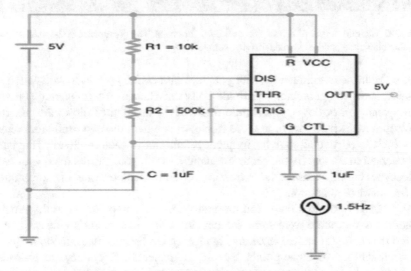

Fig. 15.5 Periodic stimulus response experiment setup. The 1.5 Hz source is obtained from a lab bench sinewave generator

oscillator circuit. This is analogous to varying the threshold of activation of a pacemaker cell.

Step 2 (Undisturbed function). Adjust your function generator to sinewave mode and set its amplitude output to zero initially. Use your oscilloscope to measure the oscillator output at pin 3. It should be still oscillating at the frequency measured in Part 1. If not, correct your circuit. Photo your output and record the frequency. This is the oscillator's "free-running" frequency or center frequency prior to any perturbation.

Step 3 (Entrainment and phase lock). Next, while observing the output pin 3 on your scope set your function generator to a frequency very near to the center frequency that you just measured slightly above or below it. Then, begin to gradually increase the function generator amplitude. You should notice that the oscillator frequency will be pulled toward the function generator frequency. It is now phase-locked. You may now vary the sinewave generator frequency. The oscillator should follow the generator frequency and remain locked to it. This is the condition of entrainment. In the heart, this condition is like the AV node following the frequency of the SA node pacemaker.

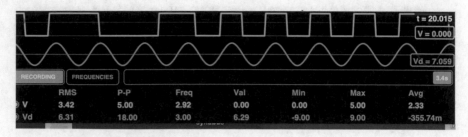

Fig. 15.6 Sample output of driven 555 oscillator for chaos. The green trace is the oscillator output. The blue trace is the sinusoidal driving source

Step 4 (Multiple entrainment). After you have successfully phase-locked your oscillator, leave the generator amplitude constant. Change the frequency generator frequency up or down. You should find that the oscillator follows the generator frequency. Experiment to find the frequencies where the two oscillators remain in lock. In particular, look for frequency multiples of each oscillator. That is, the generator at 2x and 3x the center frequency. Photo your results from your oscilloscope screen. Measure the frequencies of both oscillators and the amplitude of the sinewave generator.

Step 5 (Chaotic oscillations). You may have noticed as you changed the function generator that there were some frequencies where lock could not be maintained and the oscillator started to behave randomly. Go back to measure the frequency and amplitude of those points. Record the generator frequency for chaos and photograph the oscillator output. If you cannot observe chaos and you always have phase lock, begin to reduce the sinewave generator amplitude and try again. In this case, record the amplitude and frequency of the function generator that provided chaos. Photograph one sample of chaos that you observed. A sample of chaos is provided in Fig. 15.6 using circuit simulation. Note that frequency never stabilizes and that entrainment does not occur.

Lab Report
1. Write a summary of your experiment and discuss its relationship with earlier chapters of this book.
2. Provide oscilloscope photos that support your observations of entrainment, multiple entrainment, and chaos.
3. Provide a graph that graphically shows the various combinations of amplitude and driving frequency that resulted in entrainment and chaos.
4. Identify the characteristics of the relaxation oscillator that allow the functions that you have observed.
5. Why does a linear oscillator not exhibit phase lock?

References

DiFrancesco, D. (2010). The role of the funny current in pacemaker activity. *Circ Res, 106*(3), 434–446. https://doi.org/10.1161/CIRCRESAHA.109.208041

Hodgkin, A. L., & Huxley, A. F. (1952). A quantitative description of membrane current and its application to conduction and excitation in nerve. *J Physiol, 117*(4), 500–544. https://doi.org/10.1113/jphysiol.1952.sp004764

Correction to: Nonlinear Flow Dynamics and Chaos in a Flexible Vessel Model

Correction to:
Chapter 13 in: G. Drzewiecki, *Fundamentals of Chaos and Fractals for Cardiology*, https://doi.org/10.1007/978-3-030-88968-5_13

The chapter was inadvertently published with an incorrect symbol, on page 106, in the third paragraph, the not equal to symbol "≠" in the sentence. The vessel radius is r = square root of (A / ≠) has been replaced with "π" (Pi Symbol).

The updated version of this chapter can be found at
https://doi.org/10.1007/978-3-030-88968-5_13

Index

Printed in the United States
by Baker & Taylor Publisher Services